11 Effective Strategies for Teaching Math to Students Who Have Given Up on Learning

Student Engagement Techniques That Motivate Students With Special Needs and Ignite Excitement for Every Student in the Classroom to Be Successful

Jordan B. Smith Jr. Ed.D.

© Copyright 2023 - All rights reserved.

The content contained within this book may not be reproduced, duplicated or transmitted without direct written permission from the author or the publisher.

Under no circumstances will any blame or legal responsibility be held against the publisher, or author, for any damages, reparation, or monetary loss due to the information contained within this book, either directly or indirectly.

Legal Notice:

This book is copyright protected. It is only for personal use. You cannot amend, distribute, sell, use, quote or paraphrase any part, or the content within this book, without the consent of the author or publisher.

Disclaimer Notice:

Please note the information contained within this document is for educational and entertainment purposes only. All effort has been executed to present accurate, up to date, reliable, complete information. No warranties of any kind are declared or implied. Readers acknowledge that the author is not engaged in the rendering of legal, financial, medical or professional advice. The content within this book has been derived from various sources. Please consult a licensed professional before attempting any techniques outlined in this book.

By reading this document, the reader agrees that under no circumstances is the author responsible for any losses, direct or indirect, that are incurred as a result of the use of the information contained within this document, including, but not limited to, errors, omissions, or inaccuracies.

Table of Contents

INTRODUCTION 1

CHAPTER 1: BUILDING YOUR CLASSROOM COMMUNITY 5
   What Makes a Special Needs Classroom Different 6
   Why Does a Community Classroom Help Students? 7
   3 Things to Consider in Your Classroom Layout 8
      Furniture 9
      Layout 10
      Wall Space 11
   21 of the Best Manipulatives for the Math Classroom 12
   Classroom Plan 15

CHAPTER 2: GETTING STUDENTS REHOOKED ON MATH 19
   Ideas for Extrinsic Motivation 20
   How Students Benefit From Intrinsic Motivation 21
      What Is Intrinsic Motivation? 21
      Benefits of Intrinsic Motivation for Students 22
      How to Foster Intrinsic Motivation in Students 23
   Rehooking Your Unenthusiastic Students 24
   Getting Creative With Student Stimulation 26
      Discover Your Students' Strengths 26
      Use Positive Examples of Mathematicians With Special Needs 27
      Maximize Relationships in the Class 28
      Talk About Careers 29
      Movement and Math 29

CHAPTER 3: HOW TO OVERCOME MISTAKE ANXIETY 33
   Take Time to Teach Students About Making Mistakes 34
      Stretch Mistakes 34
      Aha Moment Mistakes 34
      Sloppy Mistakes 35
      High-Stakes Mistakes 35
      Accidental Success 36
   Develop Growth Mindsets With One Word 39
   Watch Out for Your Response 41
   Avoid the Fundamental Attribution Error 41
   Remove Stigma With the Mistake Game 43
   Stop Penalizing Practice 44
   Use the Guess-and-Check Method With Students 46
      Example 1 47

  Example 2 47
 ENCOURAGE STUDENTS TO CHECK FOR MISTAKES 48

CHAPTER 4: TEMPTING STUDENTS WITH TECHNOLOGY 51
 WHY TECHNOLOGY SHOULD BE USED IN SPECIAL EDUCATION 52
 TYPES OF TECHNOLOGY TO INCORPORATE 53
 BEST MATH APPS AND GAMES 54

CHAPTER 5: BRINGING THE REAL WORLD INTO THE CLASSROOM 59
 WHY MATH AND SPECIAL NEEDS STUDENTS REQUIRE REAL-WORLD EXAMPLES 59
 KEEPING MATH REAL 61
  Real-World Math Wall 61
  School Water Audit 62
  Pretend Restaurant 62
  Integrating Math With English and History 62
  Math Recipes 63
  Graphing Halloween Candy 63
  Other Ways to Introduce Special Needs Students to Math in the Real World 63

CHAPTER 6: PEER-BASED LEARNING 65
 WHY PEER-BASED LEARNING IS A VALUABLE STRATEGY 66
 HOW TO PREPARE STUDENTS FOR PEER-BASED LEARNING 66
 PEER-BASED LEARNING TECHNIQUES 68

CHAPTER 7: CONCRETE-TO-VISUAL-TO-ABSTRACT 71
 WHAT IS THE CONCRETE, VISUAL, ABSTRACT STRATEGY 72
 HOW TO DEVELOP A CVA UNIT 73
  Examples of CVA in Math 74

CHAPTER 8: USING THE RIGHT MATH LANGUAGE 77
 WHY SHOULD YOU BE TEACHING THE CORRECT MATH VOCABULARY 78
 HOW MUCH VOCABULARY IS NECESSARY 79
  Numbers and Operations 79
  Algebra 80
  Geometry 81
  Statistics and Probability 81
  Ratio and Proportion 82
  Exponents and Radicals 82
  Sequences and Series 82
  Domain and Range 83
  Coordinate Plane 83
  Slope 83
  Linear Regression 83
 FRAYER MODEL 84
 UNDERSTANDING WORD PROBLEMS 85

SBAC Construct-Relevant Vocabulary for Mathematics 87
Naturally Adding Vocabulary to Lessons 89

CHAPTER 9: SCHEMA-BASED INSTRUCTION 93
How Schema-Based Instruction Came About 94
Schemas for Different Word Problem Structures 95
    Addition and Subtraction 95
    Multiplication and Division 96
    Ratio and Proportion 96
Putting Schema-Based Instruction Into Action 97
    Tips for Successful Implementation 99

CHAPTER 10: RETENTION TECHNIQUES 101
The PQ4R Method 102
Retrieval Practice 104
Spaced Practice 106
The Feynman Technique 108
The Pomodoro Technique 109

CHAPTER 11: CREATING YOUR PLANS AND ACTIVITIES 113
A Step-By-Step Guide to Creating Special Education Lesson Plans 114
    Start With the Learning Target 114
    Keep Prerequisite Skills in Mind 117
    Add Scaffolding 119
    Allow for Flexibility 122
    Include Modifications and Differentiation 124
    Incorporate Technology 126
    Decide How the Material Will Be Presented 127
    Create a Guide to Complete Each Activity 129
    Decide How Progress Will Be Monitored and Assessed 130
    Include Necessary Materials 133
    Make Sure the Plan Is Aligned With Set Standards 134
Blank Lesson Plan 136
Tying It All Together 137
Where to Find Fun, Engaging Resources 138

CONCLUSION 141

ABOUT THE AUTHOR 143

GLOSSARY 145

REFERENCES 147
Image References 151

Introduction

If you can't explain something in simple terms, you don't understand it. –Richard Feynman

Mathematics is a fundamental subject in education, and for children with additional educational needs, it can be an especially challenging topic. As a teacher, you have the power to help these students unlock their potential and succeed with mathematics. However, with such a diverse range of needs in your classroom, it can be tough to know where to start. This is why I wrote this book—to help you teach math to children with special needs effectively and efficiently.

You may be feeling overwhelmed and isolated in your teaching journey. You spend countless hours planning lessons but still feel like you're not making the progress you hoped for. You're frustrated by the lack of resources and support available to you. You may even be tired of hearing teachers of typical students complain about their workload when you're dealing with so much more. I understand your struggles, and I'm here to help.

The catalyst for writing this book was seeing special needs students struggle with math concepts and recognizing the need for a better approach. The educational system is flawed, and it's not fair that your students should suffer because of it. Your students' experiences in school can influence their future outlook on life, so it's essential to make it as positive as possible. This book will give you the tools and strategies you need to do just that.

By reading this book, you will gain valuable shortcuts to make your teaching experience more efficient and enjoyable. You will learn how to use technology to your advantage, how to put math concepts into their real-world contexts, and how to engage your students in fun, interactive projects. You will be able to create a classroom environment that fosters success and supports the individual needs of each student.

I have been teaching for nearly 20 years, and during that time, I have accumulated a wealth of knowledge and experience. However, I continue to learn and grow as a teacher, constantly seeking new ways to improve my students' outcomes. The strategies and techniques I share in this book have been tried and tested, and they work.

The end result of reading this book is simple: You will become a better teacher. You will have the tools and strategies to help *all* of your students succeed in mathematics, and you will feel empowered and supported in your teaching journey. Nobody goes into teaching for the money; they do it to make a difference. This book will help you do just that.

I have the expertise and experience to guide you through the challenges of teaching math to children with additional academic support needs. I have seen firsthand the difficulties that come with teaching this subject, and I have developed strategies to overcome them. It's not easy, but with the right tools and approach, you can help your students achieve their full potential.

Before the information in this book was available, teaching math to special needs children was a daunting task. It was a process of trial and error, and many students fell through the cracks. But with the strategies and techniques outlined in this book, you can create a positive and productive learning environment for all of your students.

It's time to stop teaching on autopilot and start educating in a way that supports the unique needs of each of your students. With this book as your guide, you can make a difference in the lives of your students and help them succeed in mathematics and beyond.

Throughout this book, you will gain a deep understanding of teaching strategies from the new perspective of a special needs teacher and learn how beneficial each of the strategies can be. I have implemented these strategies and seen positive results without overwhelming my workload. Your learning will be based on proven strategies that can make a classroom inclusive to all, where neurodivergent and intellectually disabled students feel comfortable, safe, and supported. These strategies not only improve math skills but can also be transferred to other areas of learning to boost confidence and independence. You will be able to get more out of your time in the classroom, making the most of the materials and resources available. By implementing the strategies in this book, you will be able to manage your classroom more effectively and improve the educational experience for all of your students, regardless of their abilities. This book is the right choice for any teacher who wants to make a positive impact on their students' lives and become a more effective educator.

So, what are you waiting for? Let's get learning!

Chapter 1:
Building Your Classroom Community

As I walked into the high school classroom, I couldn't help but feel disappointed. The walls were completely bare, painted white, and the space felt cold and clinical. It was more like a hospital than an engaging learning environment. I couldn't imagine how any student would be inspired to learn in such a sterile setting. Gone are the days when the only objects in a classroom are the blackboard and the teacher!

What Makes a Special Needs Classroom Different

When it comes to education, one size does not fit all. This is especially true for students with learning delays and disorders who require additional support to achieve their academic and social goals. That is where special needs classrooms come in, providing a tailored learning environment to cater to the unique needs of these students.

Children with learning disabilities face a variety of challenges when it comes to absorbing information in a school environment. They may struggle to understand concepts, have difficulty communicating, or exhibit behavioral issues. These challenges can make it harder for them to keep up with the pace of a regular classroom and can lead to frustration, anxiety, and a lack of motivation.

Special education classrooms provide individualized instruction to address these challenges. Teachers in these classrooms have specialized training and expertise in working with students with academic support needs. They create individualized learning plans (ILPs) that take into account the strengths and challenges facing each student and use strategies that work best for their personal learning style.

One of the primary differences between a special needs classroom and a regular classroom is the focus on individualized instruction. In a special needs classroom, the teacher is able to work closely with each student to provide the support they need to learn and thrive. They may use alternative teaching methods, such as visual aids, hands-on activities, or assistive technology to help students understand and retain information.

Some special needs classrooms also act as complementary learning support, building on academic learning from the main class. In these cases, they may not necessarily teach content, instead helping students to apply what they have learned in the regular classroom to real-life situations. This can include developing skills such as social interaction, problem-solving, and self-advocacy. This support might also involve helping a student to meet their own learning needs—for instance, by learning how to use assistive technology or through implementing personalized study and homework strategies.

On the other hand, an inclusive classroom has more students—both with and without disabilities—all learning together. While inclusive classrooms offer many benefits—such as fostering a sense of community and promoting acceptance and understanding—they may not be suitable for all students with special needs. Students with more significant learning delays or disorders may require a more individualized approach to learning.

While this book is aimed at special needs classrooms, the techniques and strategies we will discuss can be used in both special needs and inclusive classrooms.

Why Does a Community Classroom Help Students?

A community classroom is one where students come together, feel connected, and all feel valued. It's a space in which everyone is respected and supported and where students can learn not just from their teachers but from each other. The benefits of a community classroom are many, and they extend to both students and teachers alike.

One of the key benefits of a community classroom is increased trust. When students feel like they are part of a supportive community, they are more likely to take academic risks. They are more willing to ask questions, offer opinions, and share their ideas, knowing that their classmates and teachers will be supportive and encouraging. This can lead to a more dynamic and engaged classroom, where students are actively involved in the learning process and feel empowered to take ownership of their education.

Another benefit of a community classroom is decreased behavioral issues. When students feel like they are part of a community, they are less likely to engage in disruptive or negative behavior. They feel that they are part of something bigger than themselves and are motivated to act in ways that support the goals and values of the group. This can lead to a more positive classroom environment where students are respectful and supportive of one another and where everyone feels safe and valued.

Perhaps most importantly, a community classroom helps students feel a sense of ownership over their learning. When students feel like they are part of a community, they are more invested in their education. They feel like they have a stake in their learning and are motivated to take an active role in the process. This can lead to increased engagement and academic achievement, as students are more likely to take responsibility for their own learning and work harder to achieve their goals.

But it's not just students who benefit from a community classroom. Teachers also find that they are more effective when they are part of a supportive and collaborative learning environment. They are able to build stronger relationships with their students and are better equipped to identify and respond to the unique needs of each individual. They also find that they are able to create a more engaging and dynamic classroom, where students are motivated to learn and eager to participate in the learning process.

3 Things to Consider in Your Classroom Layout

Designing a classroom environment that supports special education students is essential for their success. The physical layout of the classroom, furniture, and wall space are all important components to consider, with the most important factor being student needs.

As educators, it's natural to be drawn to aesthetically pleasing designs and trendy decor, but it's important to prioritize the needs of our students above all else. Our classrooms should be designed with the intention of supporting and enhancing learning experiences for all students, especially those with special needs.

Furniture

Teaching in a classroom with preexisting furniture can be a challenge, but there are ways to optimize the space to meet the unique needs of your students. By taking a few simple steps, you can ensure that your classroom is set up to promote student success.

When it comes to individual desks, it's helpful to use floor tape to mark off personal space. This will clearly define the area around the desk and chair, which can aid in classroom management and provide students with a sense of boundaries. It's also important to make sure that students who use assistive technology have adequate space to keep their devices accessible.

For larger tables, consider marking off personal space using colored tape on the table tops. Be sure to check the height of the tables and ensure that they are spaced appropriately to accommodate wheelchairs or other mobility aids.

Dividing your classroom into separate work areas can be useful for students with special needs, and bookshelves can serve as great moveable walls. Use them to create separate spaces for related services or group work.

If you have hard surface floors, placing tennis balls on chair legs can help minimize distracting noises and prevent sensory issues. Additionally, consider investing in a designated rug for circle time that includes personal space boundaries.

Taking the time to thoughtfully arrange your classroom can make a significant impact on the success of your students. With a few simple modifications, you can transform your space into an environment that is welcoming, supportive, and conducive to learning.

Layout

When preparing your classroom for the year, it's important to consider the layout of the space, beyond just where you want to place certain items like your classroom library, teacher desk, and computer station. You should also focus on the space between each area to ensure your students can maneuver between them, with enough room for a wheelchair or for two people to walk at the same time, in case a student with a physical disability needs assistance.

Take into account where your adult-attended spots are in relation to the independent work areas. It's recommended to place your small group table near the door so that you can easily answer the door without leaving your students. If it is age-appropriate for your students to have access to a time-out or calm-down spot, choose a quiet area of the room near an adult and away from independent workstations, sensory corners, or other stimulating activities. For teenagers and more independent learners, it may be more effective to create a solo chill-out area where they do not feel over-observed but which is still within an adult's line of sight.

Consider how your students will rotate through the room for different activities. For example, if you plan to have a life skills kitchen area, it might be distracting to place it right next to your small groups. If you have snack time right after circle time, consider placing your snack table near your rug area so that the whole class doesn't have to move across the room to transition between activities.

To plan your classroom layout, you can use chart paper and sticky notes to visualize the space and move items around until it looks good. Alternatively, you can try a free online classroom planner. Don't underestimate the importance of a well-thought-out classroom layout that fosters a conducive learning environment and promotes student success.

Wall Space

When planning how to utilize the wall space in your classroom, prioritize visual aids that will enhance your students' learning experience rather than just decorating the room. Here are a few important items that you should hang on your walls.

Daily Schedule

Having a visual schedule prominently displayed in your classroom can alleviate anxiety and help with transitions for many students, especially those on the autism spectrum. Ensure that your schedule is large enough for students to see, and use easy-to-identify pictures for nonreaders. Even if they can't tell time, they will be able to understand the order of their daily activities.

Rules and Behavior Plan

As special education is tailored to the individual needs of students, they may carry personalized behavior charts with them or keep them on their desks. However, it's still crucial to post general classroom expectations and rules that are big, easy to read, and supported by pictures. You might also consider displaying a chart for positive reinforcement alongside your rules.

Visual Supports for Learning

Your classroom's visual aids will vary depending on your grade level, but it's best to display only what your students need at the moment. Don't clutter the walls with materials that aren't relevant yet. Keep a designated spot for math aids and a separate one for reading aids to make it easier for students to locate the information they need. However, some visual aids are essential throughout the school year, including the alphabet, number line, word wall, visual cues (action and object cue cards), and a calendar with corresponding visual aids for circle time.

Once you've hung up all the essentials, you can think about decor if you still have space. But keep it simple and limit the number of distracting materials in your room. Remember, your primary focus is to create an environment that is conducive to your students' learning, not just visually pleasing.

21 of the Best Manipulatives for the Math Classroom

There are several useful tools that you can use in a variety of situations to help your special needs students learn. These tools include

- **Abacus:** helps learners visually understand and perform basic math operations like addition and subtraction.
- **Algebra tiles:** provide a hands-on way to explore algebraic concepts, particularly solving equations and factoring polynomials.

- **Cuisenaire rods:** enable learners to visualize and develop a deeper understanding of fractions, decimals, and other mathematical concepts.

- **Reflective geo mirror:** helps learners visualize and understand geometry concepts like symmetry and reflections.

- **Balance and weights:** support learning about weight and measurement while building fine motor skills.

- **Fraction tiles:** provide a concrete representation of fractions, making it easier for learners to understand and compare them.

- **Graphic calculator:** enables learners to visualize complex functions and equations and supports the development of problem-solving and critical thinking skills.

- **Centimeter grid dry-erase board:** help learners practice measurement, graphing, and problem-solving while encouraging collaboration and discussion.

- **Algebra dominoes:** provide a fun and interactive way to learn algebraic concepts, particularly solving equations and graphing functions.

- **Hands-on trigonometry proof:** supports learners in understanding and applying trigonometric concepts in a tactile way.

- **Colored blocks:** can be used for counting, sorting, pattern-making, and building spatial reasoning skills.

- **3D shapes:** enable learners to visualize and understand geometric concepts in three dimensions.

- **XY coordinate pegboards:** support learners in understanding and graphing coordinates and functions.

- **Dice:** provide an interactive and fun way to practice counting, addition, and probability.

- **Fabric tape measures:** enable learners to practice measurement and develop fine motor skills while building understanding of the real-world application of measurement.

- **Bingo chips:** can be used for counting, sorting, and pattern-making while supporting fine motor skills.

- **Pattern blocks:** encourage the exploration of geometric concepts and the development of spatial reasoning skills.

- **Numbered fidgets:** can be used to support attention and focus while building number sense and counting skills.

- **Number lines:** provide a concrete representation of number relationships and supports the development of counting, addition, and subtraction skills.

- **Geoboards:** support the development of geometric concepts and spatial reasoning skills through hands-on exploration and problem-solving.

- **Play money:** can be used to support the development of money skills and financial literacy in a fun and interactive way.

Remember that this is not an exhaustive list, you can always add more items and experiment with what works for your students.

Classroom Plan

Below, I've provided a blank classroom plan that you can use to help create an optimal learning environment for your special needs students.

With this classroom plan, you can easily organize your classroom layout, designate work areas, and make notes on additional changes you may need to make. By planning your classroom layout in advance, you can create an environment that is both functional and conducive to learning.

Take some time to consider the unique needs of your students when filling out this classroom plan. Use it as a tool to help you create a space that is inclusive, accommodating, and welcoming for all learners.

I hope this classroom plan will be helpful in creating an engaging and productive learning environment for your special needs class!

Classroom Name:

_____

Grade Level:

_____

Subject(s):

_____

Date:

_____

Room Layout

Student Desks/Tables:

Teacher's Desk:

Student Work Areas:

    Reading Area:

    Technology Area:

    Storage Area:

Wall Space

Daily Schedule:

_____

_____

_____

_____

_____

Rules and Behavior Plan:

_____

_____

_____

_____

_____

_____

Visual Supports for Learning:

- alphabet
- number line
- word wall
- visual cues

Calendar and Corresponding Visual Aids

Additional Notes

    Lighting:

    Color Scheme:

    Classroom Decorations:

    Bulletin Boards:

    Classroom Supplies:

    Classroom Library:

    Classroom Management Strategies:

    Additional Needs for Special Education Students:

Notes:

_____

_____

_____

Incorporating quotes is a fantastic way to add vibrancy to your classroom walls while imparting valuable life lessons to your students. And when it comes to math students, a well-chosen quote can be the perfect inspiration to help them dive into the subject matter.

So, let's kick off the next chapter with a quote that's just right for math students.

Chapter 2:
Getting Students Rehooked on Math

Why did the obtuse angle jump in the pool? Because it was over 90 degrees. –
Alesandra Dubin

There is always more than one way for content to sink into a student's mind (Dubin, 2022). In this chapter, we will explore strategies to foster a love for math by examining the difference between extrinsic and intrinsic motivation. We will also explore creative ideas to make math fun in the classroom. For instance, incorporating movement into math instruction can significantly benefit students with ADHD. Additionally, we will delve into building confidence in students with learning disabilities by explaining how the brain works and highlighting that everyone can learn new things when provided with the right tools.

Ideas for Extrinsic Motivation

Extrinsic motivation is an essential aspect of student learning, as it helps students stay engaged and focused on their academic goals. This type of motivation comes from external sources, and it's essential to recognize that every student is unique and motivated differently. Here are some examples of external motivators that can help students stay focused and engaged:

- **Positive reinforcements:** Praising progress rather than just results is an effective way to motivate students. Teachers can offer specific and positive feedback that recognizes the effort, creativity, and dedication students put into their work. This type of feedback helps students feel valued, appreciated, and recognized for their hard work.

- **Allowing students to choose activities:** Giving students the freedom to choose activities or tasks that interest them can help increase their motivation. When students have a say in what they learn and how they learn, they are more likely to be invested in the process.

- **Learning through play:** Incorporating play into classroom learning can be an effective way to engage students and make learning fun. Play-based learning activities can include games, simulations, and interactive tools that make the learning experience more enjoyable.

- **Rewards:** Rewards can be an effective extrinsic motivator for students, as they can help reinforce positive behavior and progress. Rewards can come in various forms, such as points, tokens, or food, and they can be given for meeting specific goals or milestones.

- **New activities:** Introducing new activities and projects can help stimulate student interest and engagement. By offering fresh and challenging activities, teachers can help students develop a sense of curiosity and excitement about learning.

- **Longer breaks:** Allowing longer breaks between tasks or activities can help students stay focused and motivated. Breaks can be used to recharge, relax, and regroup, helping students stay motivated and engaged throughout the day.

How Students Benefit From Intrinsic Motivation

As educators, we strive to help our students develop a love for learning that will stay with them throughout their lives. However, as we work to motivate our students, it's essential to recognize that motivation can come from both external and internal sources. While external motivators, such as rewards or praise, can be effective in the short term, intrinsic motivation is a more powerful, long-term driver of student success.

What Is Intrinsic Motivation?

Intrinsic motivation refers to the internal desire or drive to engage in an activity or behavior for its own sake, rather than for the benefit of external rewards or incentives. When someone is intrinsically motivated, they are motivated by the enjoyment, satisfaction, and personal fulfillment they experience from the activity itself rather than by external factors such as money, praise, or recognition. When students are intrinsically motivated, they are more likely to persist in their efforts, take on challenges, and achieve better outcomes. They are also more likely to experience a sense of autonomy and control over their learning, which can lead to greater self-confidence and self-efficacy.

Examples of activities that can be intrinsically motivating include hobbies, creative pursuits, sports, and learning new skills. Intrinsic motivation can lead to higher levels of engagement, creativity, and productivity, as individuals are motivated by their own internal desires.

Benefits of Intrinsic Motivation for Students

- **Greater engagement:** When students are intrinsically motivated, they are more engaged in their learning. They are more likely to seek out challenges and take ownership of their learning, leading to a deeper understanding and more meaningful connections.

- **Higher achievement:** Intrinsic motivation is a powerful predictor of academic achievement. Students who are intrinsically motivated are more likely to persist in their efforts, take on challenges, and achieve better outcomes.

- **Improved creativity:** Intrinsic motivation can lead to more innovative thinking and creativity. When students are motivated by their own interests and passions, they are more likely to explore and experiment with new ideas and approaches.

- **Greater resilience:** Intrinsic motivation can help students develop a growth mindset, which is essential for building resilience and overcoming setbacks. Students who are intrinsically motivated are more likely to view challenges as opportunities for growth and development.

- **Increased self-confidence:** Intrinsic motivation can lead to greater self-confidence and self-efficacy. When students feel in control of their learning and are motivated by their own interests, they are more likely to take risks and achieve success.

How to Foster Intrinsic Motivation in Students

- **Provide autonomy:** Give students the opportunity to make choices about their learning. Allow them to choose their own topics, projects, or learning activities whenever possible. This can help students feel more invested in their learning and in control of their own educational journey.

- **Offer opportunities for mastery:** Provide students with opportunities to develop their skills and knowledge in meaningful ways. When students experience success and feel a sense of competence, they are more likely to be motivated to continue learning.

- **Connect learning to real-world applications:** Help students understand the practical applications of what they are learning. When students can see how their learning applies to real-world situations, they are more likely to be motivated to engage in the learning process. This is especially useful in mathematics as many of the concepts can seem "abstract" and difficult to understand. How often have you heard a student say they'll "never use this in real life"? Most say this because they have no idea how or why mathematical concepts are used, especially when you start getting into higher-level topics like calculus.

- **Encourage curiosity:** Encourage students to ask questions, explore new ideas, and take on challenges. When students are motivated by their own curiosity and interests, they are more likely to be engaged in the learning process.

- **Provide feedback and encouragement:** Provide specific and meaningful feedback that recognizes the effort, creativity, and dedication students put into their work. Celebrate their successes, and provide support and encouragement when they face challenges.

Rehooking Your Unenthusiastic Students

While there are several approaches to increasing student motivation, one often-overlooked aspect is the teacher's behavior in the classroom. Let's explore how teachers can increase student motivation through their own actions, specifically in the context of mathematics:

- **Discovering voids in learning:** One of the most effective ways to increase motivation in math is to identify the gaps in students' understanding. By identifying areas where students may be struggling, teachers can provide targeted support to help them build their confidence and understanding.

- **Show sequential achievement:** Students need to see progress and understand how their efforts lead to success. By breaking down complex concepts into smaller, more manageable parts, teachers can help students see the sequential achievements that lead to mastery.

- **Discover a pattern:** Humans are wired to detect patterns, and the pupils in a math classroom are no exception. By highlighting patterns in mathematical concepts, teachers can help students understand the underlying structure and logic of mathematical ideas.

- **Present a challenge:** Challenges can be highly motivating, especially for students who are confident in their abilities. By introducing challenges that are within reach but require effort and creativity to solve, teachers can help students feel a sense of accomplishment and pride in their work.

- **Introduce a "gee whiz" math result:** Math can be full of surprises and unexpected results. By introducing students to math results that defy intuition, teachers can pique their curiosity and inspire them to explore further.

- **Use recreational math:** Puzzles, paradoxes, and other recreational math activities can be highly engaging and motivating for students. By incorporating these types of activities into the classroom, teachers can help students see math as a fun and exciting subject.

- **Tell a pertinent story:** Stories can be a powerful way to help students connect with mathematical concepts on a personal level. By telling stories that illustrate the real-world applications of mathematical concepts, teachers can help students see the relevance and importance of math in their lives.

- **Introduce mathematical curiosities:** The Fibonacci series, the golden ratio, and other mathematical curiosities can be fascinating and thought-provoking for students. By introducing these types of curiosities, teachers can help students see the beauty and elegance of mathematical ideas.

Increasing student motivation for math is not just about understanding the internal and external factors that influence student behavior. Teachers can play a significant role in increasing student motivation by being intentional in their behavior and actions. By following the guidelines above, teachers can create a positive and motivating learning environment that encourages students to explore and engage with math.

Getting Creative With Student Stimulation

As an educator, it's essential to recognize that not all students learn in the same way. While traditional teaching methods have their place, it's important to get creative with student stimulation to engage and inspire all learners. Let's explore some unique and creative ideas for stimulating student learning.

Discover Your Students' Strengths

As an educator, it's important to recognize that every student has unique strengths, talents, and abilities. By understanding each student's individual learning style and preferences, you can tailor your teaching methods to meet their specific needs. This can not only help to increase their engagement and motivation but can also lead to improved academic performance.

One way to identify your students' strengths is to regularly assess their learning styles and preferences. This can be done through a variety of methods, such as surveys, classroom observations, and one-on-one discussions. By gathering this information, you can better understand how each student learns best and can tailor your lesson plans accordingly.

For example, if you have a student who has a talent for visual learning, incorporating visual aids and graphics into your lessons can be an effective way to engage and stimulate their learning. You can use tools such as diagrams, charts, and infographics to help them visualize complex concepts and better understand the material. Similarly, if you have a student who is a natural auditory learner, incorporating lectures, discussions, and group discussion activities can help them stay engaged and retain information.

By tailoring your teaching methods to each student's individual strengths, you can create a more engaging and effective learning environment. This can not only help students stay motivated and interested in the subject matter but can also help them develop confidence in their abilities and reach their full potential. Additionally, it can promote a sense of inclusivity in the classroom, where every student feels valued and respected for their unique skills and talents.

Use Positive Examples of Mathematicians With Special Needs

Math can be a challenging subject for students with special needs, and this can often lead to feelings of frustration, discouragement, and a lack of motivation. However, it's important to remember that there are many examples of successful mathematicians who have overcome similar challenges. By sharing positive examples of mathematicians with special needs, you can inspire your students and show them that success is possible.

There are many well-known mathematicians who were neurodivergent such as Albert Einstein, who is believed to have had dyslexia and ADHD, and Temple Grandin, who has autism. These individuals faced many obstacles in their lives but were able to overcome them through perseverance, hard work, and creative problem-solving skills. Students with dyscalculia who might perceive themselves to be more creative than mathematical in nature may, likewise, be motivated by examples such as Leonardo da Vinci and monastic manuscript illuminators, who used mathematical concepts to facilitate their world-renowned art.

By sharing these examples with your students, you can help them to see that they are not alone in their struggles, there are ways to overcome their difficulties, and their success is not determined by their natural strengths and challenges. You can also use these examples to demonstrate the importance of developing coping strategies and utilizing resources, such as assistive technology, tutoring, and specialized instruction.

Promote a growth mindset among your students by emphasizing that intelligence and ability are not fixed traits, but instead, they can be developed and improved through effort and practice. By encouraging your students to adopt a growth mindset and to view challenges as opportunities for growth, you can help to build their confidence and motivation.

Ultimately, by sharing positive examples of successful mathematicians with special needs, you can inspire your students and help them to see that with hard work and perseverance, they too can achieve great things. This can be a powerful tool in fostering a sense of motivation and engagement in math and other subjects.

Maximize Relationships in the Class

Positive relationships between teachers and students, as well as among peers, are essential to fostering a positive and engaging learning environment. Students who feel connected and supported by their teachers and peers are more likely to feel motivated and engaged in their learning. As a teacher, it's important to take the time to get to know your students on a personal level and to create opportunities for peer-to-peer relationships to develop.

One way to build positive relationships with your students is to take an interest in their lives outside of the classroom. Ask them about their hobbies, interests, and goals, and find ways to incorporate these into your teaching. For example, if you have a student who is interested in music, you could use musical examples to teach mathematical concepts such as rhythm and pattern recognition.

Collaborative activities and group projects are also effective ways to encourage positive relationships among peers. By working together on a shared goal, students learn to communicate effectively, build trust, and develop a sense of accountability to one another. Additionally, working in groups can help to break down barriers between students and foster a sense of inclusivity and belonging.

Talk About Careers

Another way to promote motivation and engagement in math is to discuss potential career paths in the field. By showing students the real-world applications of mathematical concepts and the various career opportunities available to those with a strong math background, you can help them see the relevance and importance of the subject matter. For example, you could invite guest speakers who work in math-related fields to talk to your class or arrange field trips to companies or organizations that use math in their work.

Movement and Math

Incorporating movement into math lessons is a great way to engage students who may have had negative experiences with math in the past and to provide a more inclusive and dynamic learning environment. Movement-based activities and games not only make learning math more enjoyable, but they can also improve students' physical and mental health.

Movement-based activities can range from simple exercises such as stretching or taking a quick walk around the classroom to more elaborate games and challenges that require physical movement. For example, a game of "Math Dodgeball" can involve students answering math questions while dodging foam balls, or a game of "Math Charades" can involve students acting out mathematical concepts or equations or one student graphing an equation using beanbags and a grid on the floor while the rest try to work out what the equation must be.

Research has shown that incorporating movement into learning can be particularly beneficial for students who have trouble staying focused or engaged while sitting at a desk. Movement-based activities can help students release pent-up energy, reduce stress, and improve their overall mood and well-being. Additionally, movement-based learning can also improve cognitive function and memory retention (*Math & Movement*, 2015).

By incorporating movement into math lessons, teachers can create a more inclusive and dynamic learning environment that meets the needs of a wider range of students. Students who may have previously struggled with math due to a lack of engagement or interest can benefit greatly from this approach, while also improving their physical and mental health.

As a math teacher, I am always looking for new and creative ways to engage my students in the subject matter. One day, I had an idea that I thought would really capture their imaginations: I decided to create a grocery store in class!

I brought in a few additional cartons of snacks and asked each of my students to take out their own snacks as well. We set up a few tables and arranged the snacks as they would be in a real grocery store.

Then, I explained to my students that the activity was to take a recipe, look at the amounts of each ingredient they needed, and decide how much money was needed to purchase those items, taking into consideration any discounts that might be available.

I could see the excitement building on my students' faces as they began to work together in groups, discussing the prices of each item and calculating how much they would need to purchase all of the necessary ingredients for the recipe. As the activity progressed, I noticed that my students were really starting to understand how math concepts like addition, subtraction, and percentages could be applied in real-world situations. And, they were having fun while doing it!

At the end of the activity, we all gathered together to share our calculations and discuss any challenges we faced along the way. My students were eager to share what they had learned and how they had applied their math skills in a practical and fun way.

Overall, the grocery store activity was a huge success, and I could see that my students were more engaged and motivated in their math studies as a result. I felt proud to have created an experience that not only taught them valuable skills but also allowed them to enjoy the learning process.

It's refreshing to see a class of students actually excited to learn a challenging subject. But what happens when you ask a question and a look of fear passes over all of their faces? Even if they know the answer, there may still be a fear of getting it wrong! In the next chapter, we will discuss how to help your students overcome mistake anxiety.

Chapter 3:
How to Overcome Mistake Anxiety

Kids are supposed to make mistakes. That's why we have erasers! –Unknown

People who have not had the best time with math might feel extremely nervous about even trying to solve problems for fear of making a mistake. This chapter will provide numerous ideas for teachers to create an environment where mistakes and failure are a normal part of life.

Take Time to Teach Students About Making Mistakes

Mistakes are often viewed in a negative light or as something to be avoided, but the truth is that mistakes are a part of learning, and while as teachers we may understand this, oftentimes for our students, this simple fact won't be so obvious. Students will often put too much pressure on themselves, and this can lead to anxiety and disappointment when they do eventually make mistakes. So, it is a good idea to explain to our students that mistakes are a normal part of learning and that what's important is learning from our mistakes and not repeating them. You can do this by breaking mistakes down into four categories.

Stretch Mistakes

Stretch mistakes are made by attempting something beyond a person's current abilities. These mistakes challenge us to learn new knowledge and skills. Encourage students to start with easier tasks and gradually work their way up, or try different learning strategies if these mistakes persist.

Aha Moment Mistakes

Mistakes that cause students to realize they have done something wrong or ineffective are *aha moments*. These mistakes provide insight, and the correct course of action becomes apparent. Encouraging students to reflect on their mistakes, learn from them, and make adjustments accordingly can turn more mistakes into aha moments and ultimately build students' self-confidence.

Sloppy Mistakes

Mistakes made by losing focus while doing something that we know well are sloppy mistakes. These instances can remind us to focus on the task at hand. Encourage students to use the Pomodoro technique to stay focused and to check their work before submitting it. The Pomodoro technique is a time management method developed by Francesco Cirillo in the late 1980s. It's a simple yet effective technique that can help people improve their productivity and focus. The technique involves breaking down work into 25-minute intervals, separated by short rest periods. Here's how it works:

1. Choose a task to work on.
2. Set a timer for 25 minutes.
3. Work on the task for the entire 25 minutes without interruption.
4. When the timer goes off, take a short break (5–10 minutes).
5. After the break, start another 25-minute work session.
6. After four work sessions, take a longer break (15–30 minutes).

The idea behind the Pomodoro technique is to help people work in short, focused bursts, allowing them to concentrate more effectively and avoid distractions. By breaking work into smaller chunks, it can be less overwhelming and easier to manage. The short breaks also provide an opportunity to recharge and refocus, which can help improve overall productivity.

High-Stakes Mistakes

These are mistakes made in high-pressure situations where the stakes are high. This can be an exam or an important test where a student draws a "blank." Encourage students to use the process of elimination instead of random guessing and to make informed judgments using their prior knowledge; taking deep breaths can help, in some cases, to both clear the mind and relieve anxiety in the moment.

Accidental Success

Some of the world's most significant discoveries and inventions were the result of a mistake. As educators, it helps our students understand that making mistakes is a natural part of the learning process. Here are some examples that you can give of things that have been made by mistake.

Penicillin

Sir Alexander Fleming, a Scottish biologist and pharmacologist, made one of the most significant discoveries in the history of medicine when he accidentally discovered penicillin. In 1928, while working at St. Mary's Hospital in London, Fleming was researching ways to treat bacterial infections. At that time, bacterial infections were a significant cause of illness and death, and doctors had limited options to treat them.

One day, Fleming noticed that a petri dish containing *Staphylococcus* bacteria had been contaminated with a mold. He initially intended to discard the contaminated dish but instead decided to examine it more closely. To his surprise, he found that the bacteria surrounding the mold had been destroyed, while the bacteria farther away from the mold were still growing.

Intrigued by this observation, Fleming conducted further experiments to confirm his findings. He found that the mold—which he identified as a strain of *Penicillium*—produced a substance that could kill a wide range of harmful bacteria, including those that caused pneumonia, meningitis, and other serious infections.

Fleming published his findings in 1929, but his discovery did not immediately lead to the widespread use of penicillin. It was not until the 1940s, during World War II, that researchers in the United States, led by Howard Florey and Ernst Chain, were able to mass-produce penicillin, making it available to treat soldiers wounded in battle. Penicillin was then made available for civilian use, and it revolutionized the field of medicine, allowing doctors to effectively treat previously deadly bacterial infections.

Fleming's discovery of penicillin has been called one of the most important medical discoveries of the 20th century. It has saved millions of lives and has been used to treat a wide range of bacterial infections. Fleming was awarded the Nobel Prize in Physiology or Medicine in 1945, along with Florey and Chain, for their work on penicillin.

Chocolate Chip Cookies

Ruth Wakefield was a skilled cook and the owner of the Toll House Inn, a popular restaurant located in Whitman, Massachusetts. Her menu was filled with delicious dishes, but one dessert stood out above the rest: her chocolate chip cookies.

It all started when Ruth decided to make a batch of her famous Butter Drop Do cookies, a recipe she had used for years. As she began mixing the ingredients, she realized that she was out of baker's chocolate, a key ingredient in the recipe. Without it, her cookies would be missing their signature chocolatey goodness.

Not wanting to disappoint her guests, Ruth quickly improvised. She grabbed a bar of Nestle's semi-sweet chocolate and chopped it into small pieces, thinking that they would melt and spread throughout the cookie dough like baker's chocolate would.

To her surprise, the chocolate pieces did not melt as she had expected. Instead, they held their shape and added a delightful texture and taste to the cookies. Ruth's guests loved them, and the Toll House Inn's chocolate chip cookies quickly became a hit.

Word of Ruth's delicious cookies spread quickly, and soon, she was receiving orders from all over the country. In 1939, she struck a deal with Nestlé, which agreed to provide her with all the chocolate she needed in exchange for the rights to use her recipe and the Toll House name.

Today, the Toll House chocolate chip cookie is one of America's most beloved desserts, enjoyed by millions of people every year. And it all started with Ruth Wakefield's creative improvisation in the kitchen.

Post-it Notes

Spencer Silver was a scientist at 3M, a company known for developing innovative products. In the late 1960s, he was working on a project to create a strong adhesive that could be used in the aerospace industry.

Silver spent countless hours in the lab, experimenting with different formulas and combinations of materials. But no matter how hard he tried, he couldn't seem to create an adhesive that met the strength requirements of the aerospace industry.

In his frustration, Silver began to experiment with a different approach. He decided to create an adhesive that was *weaker* than the current standard or anything he had been trying to achieve thus far. This way, it wouldn't damage the surfaces it was applied to and could be easily removed and repositioned.

Silver succeeded in creating a weak adhesive, but at the time, he wasn't sure what use it could possibly have. He presented his findings to his colleagues at 3M, hoping that someone else might see a potential application.

One of Silver's colleagues, Art Fry, was also struggling with a problem. As a choir singer, he was constantly trying to find a way to mark his place in his hymnal without damaging the pages. He had tried using bits of paper, but they would often fall out or get lost.

When Fry heard about Silver's weak adhesive, he had an idea. He used the adhesive to attach bits of paper to his hymnal and found that it worked perfectly. The paper stayed in place but could easily be removed and repositioned without damaging the pages.

Fry and Silver worked together to refine the adhesive, eventually creating what we now know as Post-it Notes. Today, Post-it Notes are used all over the world, in offices, homes, and schools, for everything from jotting down reminders to leaving messages for friends and coworkers.

Spencer Silver's accidental decision to invent a weak adhesive may not have been what he set out to achieve, but it ended up revolutionizing the way we communicate and organize our thoughts. It just goes to show that sometimes, the best ideas come from unexpected places.

Develop Growth Mindsets With One Word

A fixed mindset is one wherein students believe that they can't learn new things and that they can't develop new skills and talents. This is common in those who have struggled and given up. A growth mindset, on the other hand, is one that prompts students to recognize that mistakes are part of learning and that they are able to build on their abilities. Let's explore five techniques that can help special needs students develop a growth mindset:

- **Start slow and use small steps:** Break down tasks into smaller, achievable goals that can be reached gradually. Encourage students to focus on progress not perfection. Celebrate small victories, and acknowledge their efforts.

- **Enforce positive behaviors:** Praise and reward positive behaviors, such as persistence, effort, and resilience. Avoid labeling students, and instead, focus on their strengths and potential. Create a positive and supportive learning environment that fosters growth.

- **Anticipate negative behaviors:** Identify potential triggers or challenges that may lead to negative behaviors. Develop strategies and plans to address and overcome these challenges. Encourage students to communicate their feelings and concerns.

- **Share stories of other people's success:** Highlight stories of successful individuals with similar challenges or disabilities. Encourage students to learn from these stories and see challenges as opportunities for growth. Foster a sense of community and support among students.

- **Check in often:** Regularly check in with students to monitor their progress and address any challenges or concerns. Provide feedback and support to help students stay on track. Encourage students to reflect on their learning and growth.

A really simple way to encourage students is to add the word *yet* to the end of their sentences: "I can't do this problem" vs "I can't do this problem *yet*."

Watch Out for Your Response

How you respond to your students' mistakes and questions can have a massive impact on their self-esteem and how they view learning. Never respond with just "no." It's important to explain what is wrong so that they can learn. Also, while explaining where a student went wrong, you should point out any areas where they were correct.

Acknowledging your students' thought processes means valuing the mental effort they use to arrive at an answer. It's important for you to recognize that students do not arrive at an answer in a vacuum. They use their own reasoning, background knowledge, and experiences to form a response to a question. Therefore, you should focus on understanding and appreciating their thought process rather than just focusing on the final answer.

Correcting a student's thought process is often more beneficial than simply correcting their answer. When you focus on correcting the answer only, you miss an opportunity to help students understand how they arrived at the incorrect answer in the first place. By acknowledging and correcting a student's thought process, though, you can help your students understand the reasoning behind the correct answer and improve their future problem-solving skills.

Avoid the Fundamental Attribution Error

A fundamental attribution error is a psychological phenomenon wherein people tend to attribute the actions and behaviors of others to their personality traits or character rather than considering external situational factors—for example, thinking that someone did badly in an exam because they are lazy and didn't study, instead of considering that they may be tired or distracted. It's common for people to judge others based on their actions without fully thinking about the context in which those actions were performed. This can lead to misunderstandings, conflicts, and miscommunication, both in personal and professional relationships. This is where the phrase "putting yourself in someone else's shoes" can be particularly useful.

In the classroom, the fundamental attribution error can have a significant impact on students' learning and academic performance. When one student assumes that another is "stupid" for not knowing the answer to a question, they are making the mistake of attributing the other student's lack of knowledge to their intelligence or lack thereof rather than considering external factors that may be contributing to the situation. For example, the student may not have understood the material because the teacher did not explain it clearly or did not provide enough practice and feedback.

Jumping to this type of conclusion about others also causes students to berate themselves when they make mistakes. A student may blame a bad grade on their lack of math skills or intelligence rather than considering other factors such as not studying hard enough or not seeking help when needed. This can lead to a lack of motivation, increasing self-doubt, and a negative attitude toward learning.

To overcome the fundamental attribution error, consider the external factors that may be contributing to your students' behavior or actions. In the classroom, you can help your students by creating a supportive and inclusive learning environment, where students feel comfortable asking questions and seeking help. You can also provide clear and concise explanations, examples, and practice opportunities to help students develop a deep understanding of the material.

Get students to reflect on their own learning process and identify areas where they need improvement. By focusing on their own efforts and strategies, students can take ownership of their learning and develop a growth mindset, which can lead to improved academic performance and motivation.

Remove Stigma With the Mistake Game

The Mistake Game is an ideal way to get students comfortable with making errors. One student will make an intentional mistake in math while other students ask good questions about the mistake rather than just pointing out mistakes. So, how do you get your students to play the mistake game? Simple, just follow the five steps below!

1. To begin the activity, each student should work on their own to complete a set of math problems, usually consisting of eight or nine questions.

2. The students will then be grouped, and each group will be assigned one of the math problems to present to the entire class.

3. Within their groups, the students will share their individual solutions and deliberate on which solution they would like to present to the class using a whiteboard or the ELMO projector.

4. Here's the crucial step! The group must intentionally make a mistake in their solution. They can either choose a mistake made by one of their group members—which leads to a discussion of who had the "best mistake"—or they can decide on a mistake that other students might make. They are allowed to make as many unintentional mistakes as they want.

5. As each group presents their solutions, the rest of the class is expected to listen and attempt to find the mistakes made by the presenting group. Instead of simply pointing out the mistake when they identify it, they must ask a question to get the group

to admit their error, such as "Why did you...?" or "Can you explain how you did...?"

The Mistake Game has been proven to be incredibly effective in math classes. Students enjoy the process of finding mistakes made by their peers and sharing them with the class. It also enables them to learn how to ask pertinent questions when they spot a mistake instead of merely pointing it out. This game serves as an excellent review after a test, as students love to share their "best mistakes" and see that other students made similar errors. My students consistently ask to play the Mistake Game, even when it isn't technically considered a game! Sometimes, we're just correcting homework together, but the game creates a fun and safe environment in which to admit to errors, and students love that.

Stop Penalizing Practice

Grading using a percentage value on a student's work is the most commonly used form of evaluating a student's work; however, while grading can provide feedback and motivation to students, it can also have negative consequences, particularly for students facing additional academic challenges. Grading practices can reinforce negative aspects of students' math abilities, leading to feelings of shame, fear, and anxiety.

Traditional grading practices in math often focus on the final product—the correct answer—rather than the process or effort put into solving the problem. This approach can lead students to feel as though they are not "good" at math, even if they have put in significant effort, perhaps even using the correct mathematical concepts but accidentally inverting a fraction or misreading a numeral or sign. Students who struggle with their grades may feel like they are not intelligent, in turn leading to feelings of shame and inadequacy. Students who receive low grades may develop a fear of math, feeling like they are not capable of understanding the subject matter. This can lead to a negative cycle, where students avoid math or become disengaged from the subject.

Traditional grading practices may also reinforce the idea that math is a fixed ability rather than a skill that can be developed with practice and effort. Students who receive low grades may believe that they are simply not good at math rather than recognizing that they may need additional support or practice tailored to their learning style. This can lead to a fixed mindset, wherein students believe that their abilities are predetermined and cannot be changed.

There are a couple of alternative grading practices that can help mitigate these negative effects of traditional grading practices in math. One approach is to provide feedback that focuses on the process rather than the final answer. This can help students understand where they went wrong in their problem-solving process and give them guidance on how to improve their skills.

You can also try using formative assessments, which are assessments designed to provide feedback on students' understanding of a concept while they are still in the process of learning. Formative assessments can help teachers identify areas in which students are struggling and enable us to provide targeted support to help them improve their skills. This can be particularly helpful for students who may be struggling with math, as it can give them the opportunity to receive additional support before they become discouraged or disengaged.

Standards-based grading focuses on students' mastery of specific skills or standards rather than their overall performance. This approach can help students understand that their abilities are not fixed and that they can improve with effort and practice. Standards-based grading can provide a clear roadmap for students to follow, helping them understand what they need to do to achieve mastery of a particular skill. Which grading methods you use will depend on your teaching style and your students' needs, but don't stick to only one method, try out several of them or even use combinations of them to see which works best in your class.

Remember the point of a test is to find out what a student knows and doesn't know so that they can focus on improving their overall knowledge.

Use the Guess-and-Check Method With Students

The guess-and-check method is a problem-solving strategy that involves making an initial guess, testing it to see if it works, and then refining the guess until a solution is found. This method is particularly useful when the problem at hand does not have a clear-cut solution. This can be a good way for your students to figure out a problem on their own. The steps your students need to follow are

1. **Understand the problem:** Read the problem carefully to understand what is being asked. Identify the variables involved and what needs to be solved.

2. **Make an initial guess:** Based on your understanding of the problem, make an initial guess. This guess does not need to be accurate, but it should provide a starting point for further refinement.

3. **Test the guess:** Use the guessed solution to see if it works. If it does, the problem is solved. If it does not, move on to the next step.

4. **Refine the guess:** Based on the information gained from testing the guess, refine it to get closer to the solution. This may involve

adjusting the guess, making an educated guess based on patterns or logic, or trying a different approach.

5. **Repeat steps 3 and 4:** Continue testing and refining the guess until a solution is found.

Let's go through some examples of how the guess-and-check method can be used in math.

Example 1

Find the value of x that satisfies the equation $x^2 - 5x + 6 = 0$.

Step 1: Understand the problem. We need to find the value of x that satisfies the equation.

Step 2: Make an initial guess. Let x = 1.

Step 3: Test the guess. Plugging x = 1 into the equation, $1 - 5 + 6 = 2 \neq 0$. Therefore, the guess must be refined.

Step 4: Refine the guess. We can try increasing the value of x. Let x = 2.

Step 5: Repeat steps 3 and 4. Testing the refined guess by plugging x = 2 into the equation, $4 - 10 + 6 = 0$. Therefore, 2 is a solution and a zero of this quadratic. No further refinement is needed.

Example 2

A number is equal to the sum of its digits plus 18. What is the number?

Step 1: Understand the problem. We need to find a number that is equal to the sum of its digits plus 18.

Step 2: Make an initial guess. Let the number be 35.

Step 3: Test the guess. 35 does not equal the sum of its digits plus 18; 3 + 5 + 18 = 36. So, we need to refine our guess.

Step 4: Refine the guess. We can try changing the number. Let the number be 28.

Step 5: Repeat. Testing the refined guess, 28 equals the sum of its digits plus 18; 2 + 8 + 18 = 28.

Encourage Students to Check for Mistakes

Allowing your students to check their own work gives them the chance to look for errors in the process rather than just whether the solution is right or wrong, this can be particularly helpful for math concepts. A good method you can use is the TOTE method. Students can do this on their own and can use it in more subjects than just math, to apply it follow these steps:

1. **Test** the students' knowledge.

2. **Operate**—study only the parts that they got wrong.

3. **Test** them again.

4. **Exit** the topic once they get all the questions right.

One day, during a class on multiplication, I made a simple mistake. I was trying to demonstrate how to solve a problem, but I mixed up two of the numbers, and my answer was incorrect. The students immediately noticed the error and started to giggle.

At that moment, I could have gotten defensive and tried to explain it away, but instead, I took a deep breath and said, "You know what, you're right. I made a mistake. Thank you for pointing it out." I then went on to show the class how to correct the mistake and solve the problem correctly.

The students were surprised by my response. They had never seen a teacher admit to making a mistake before. But they also felt validated and empowered because I had shown them that making mistakes was a natural part of learning.

From then on, I made it a point to show my students that making mistakes was nothing to be embarrassed about. I encouraged them to share their mistakes with the class so that we could learn from them together. I noticed that the students became more confident and engaged in class, and their grades began to improve. My approach not only helped my students become better mathematicians but also taught them an important life lesson about the value of honesty and vulnerability.

Our next strategy is one that many students can't resist, which also helps them to overcome the fear of making mistakes. As technology is such an important part of students' lives, it makes sense to incorporate it and take advantage of its benefits.

Chapter 4:
Tempting Students With Technology

Why do computer scientists get Halloween and Christmas mixed up? Because Oct 31 = Dec 25. –Unknown

Not all learning can revolve around technology, but there are some awesome apps and games that can help bring concepts together and allow children to improve their math skills in a different way. As technology is such an important part of our world, math with technology is like learning two life skills in one. In this chapter, we will take a look at some of the ways that you can integrate technology into your class.

Why Technology Should Be Used in Special Education

Until fairly recently, most disabled and neurodivergent students were either excluded or at a severe disadvantage in the classroom. However, with the advent of new technologies, this is no longer the case. One of the most noticeable ways that technology can help is by simplifying communication between students and teachers. For students who struggle with speech, technology provides tools such as speech-to-text and text-to-speech software, which allow them to communicate more easily and effectively. This improves their ability to participate in classroom discussions, express their ideas and feelings, and engage with their peers and you.

Technology also enhances engagement among students by offering interactive and multisensory learning experiences that cater to different learning styles. Students can use virtual reality tools to explore the ways geometry presents itself in so much of the world or engage with interactive whiteboards to solve math problems in a visual and engaging way. This approach to learning ensures that students are fully engaged and motivated, making it easier for them to grasp complex concepts and develop their skills.

Assistive technologies such as speech recognition software, eye-tracking devices, and alternative input devices, allow students with special needs to perform tasks without additional human assistance. This can improve their self-confidence and self-esteem, as they are able to complete tasks independently.

Students may benefit from using calming apps or other digital tools that help them manage their emotions and anxiety levels. This can be particularly helpful for students with autism who may find it difficult to manage their emotions in social or sensorily stimulating situations.

By using technology in the classroom, students are exposed to new advancements and digital tools that are becoming increasingly important in many industries. This ensures that they are better prepared for the future and have the necessary skills to succeed in a rapidly changing world.

Types of Technology to Incorporate

Assistive technologies help to address different areas of learning disabilities so that students can capitalize on their strengths and improve their skill deficits. Though few of these technological options are geared specifically toward improving mathematical performance, having a wide range of technology can be used to support other areas of math; for instance, technology that helps with reading and writing can also help with learning math vocabulary and word problems. In your classroom, be sure to include a list of the kinds of assistive technology tools available so that your students can feel a level of autonomy over choosing how to accommodate their own needs. Most of these technologies will require the use of a tablet or computer, though this has the advantage of being able to use several different kinds of tech with the same hardware. Some of the technology that you should consider incorporating into your class include

- **Text-to-speech software:** converts written text into spoken words, which can help students who struggle with reading comprehension or have visual impairments.

- **Speech-to-text software:** converts spoken words into written text, which can help students who struggle with writing or have physical impairments.

- **Math notation software:** helps students to create and manipulate mathematical equations and symbols, which can be particularly useful for students with dyscalculia.

- **Graphic organizers:** helps students to organize their thoughts and ideas visually, which can aid in reading comprehension, writing, and problem-solving.

- **Electronic timers and reminders:** help students to manage their time more effectively and stay on task.

- **Digital recorders:** allows students to record lectures, instructions, or other information to review later, which can help with retention and comprehension.

- **Assistive listening devices:** amplifies sound and reduces background noise, which can be helpful for students with hearing impairments.

- **Alternative keyboards and mice:** offer alternative input methods for students who have physical disabilities or struggle with fine motor skills.

This isn't an exhaustive list, and what will work in your class will depend on the needs and abilities of your students.

Best Math Apps and Games

Math games can be a great way for students to both learn and associate a positive attitude toward math. Some of the best apps include

- **Achieve 3000:** *Achieve 3000* is an adaptive math program that personalizes learning for each student. It uses interactive video lessons, practice problems, and real-time feedback to help students master math concepts.

- **Muzology:** *Muzology* is an educational technology company that creates music-based videos to help students learn and retain information in subjects such as math, science, and history. The videos combine catchy tunes with educational content to make learning fun and engaging. *Muzology*'s videos are aligned with Common Core standards and are used by schools, teachers, and students across the United States.

- **Let's Go Learn:** *Let's Go Learn* is an educational technology company that provides online assessments and personalized learning solutions for students. Their assessments cover a wide range of subjects—including reading, math, and language proficiency—and are used by schools and educators to identify students' strengths and weaknesses and develop individualized learning plans. *Let's Go Learn* also offers instructional materials and tools for teachers to help them support student learning and growth. Their solutions are designed to be adaptive, meaning they adjust to each student's needs and progress to provide tailored support and instruction.

- **Mangahigh:** *Mangahigh* is a game-based math learning platform that engages students in math concepts through fun and interactive games. It covers a wide range of topics and has a comprehensive reporting system for teachers.

- **Woot Math:** *Woot Math* is an online math program that focuses on conceptual understanding and problem-solving. It has adaptive scaffolding that adjusts to each student's level and provides targeted feedback.

- **ALEKS:** *ALEKS* is an adaptive learning program that assesses each student's math knowledge and then creates a personalized learning plan to help them fill in knowledge gaps and advance to higher levels.

- **Mathspace:** *Mathspace* is an online math program that uses adaptive technology to create a personalized learning experience for each student. It offers interactive lessons, videos, and practice problems.

- **Knowre:** *Knowre* is an adaptive math program that provides students with personalized learning paths based on their strengths and weaknesses. It uses interactive video lessons and adaptive practice problems to help students master math concepts.

- **CK-12:** *CK-12* is an online resource that provides free math textbooks, videos, and interactive activities for high school students. It covers a wide range of math topics and allows students to learn at their own pace.

- **PhET Interactive Simulations: Math:** *PhET Interactive Simulations* is a collection of interactive simulations for math topics such as algebra, geometry, and calculus. It allows students to explore and visualize math concepts in a fun and interactive way.

- **CueThink:** *CueThink* is a collaborative math learning platform that encourages students to solve problems together. It offers real-world problem-solving challenges and peer-to-peer feedback.

- **Desmos:** *Desmos* is an online graphing calculator that allows students to explore and visualize math concepts. It offers a variety of features such as sliders, tables, and interactive graphs.

- **Mathalicious:** *Mathalicious* offers real-world math lessons that help students make connections between math and their daily lives. It offers interactive lessons and videos.

- **Virtual Nerd:** *Virtual Nerd* offers video tutorials on a wide range of math topics for high school students. It creates personalized learning paths based on each student's needs.

- **Brilliant:** *Brilliant* offers a wide range of math and science courses for high school students. These are made engaging through the use of interactive lessons, quizzes, and challenges, all of which help students master math concepts.

- **GeoGebra:** *GeoGebra* is an online platform that offers math tools such as a graphing calculator, geometry tools, and a spreadsheet. It allows students to visualize and explore math concepts.

- **ExploreLearning Gizmos: Mathematics Grades 9-12:** *ExploreLearning Gizmos* is a collection of interactive simulations for high school math topics. It enables students to explore and visualize math concepts in a fun and interactive way.

Most of these are aimed at students in high school but can be adapted for use with lower grades as well. Some are free or offer free trials, so it's worth checking each of them out to see which ones will best suit your classroom.

Sometimes, the thought of creating games through different apps can seem like an exhausting and challenging one, but remember, once they are prepared they can be used time and time again, even shared with other math teachers.

Technology is a perfect way to engage students because it uses the real world to put learning into context. When students struggle with difficult concepts, teachers need to find ways to show them how their learning is relatable to the world outside the classroom. In the next chapter, we will explore how you can bring the real world into the classroom.

Chapter 5:
Bringing the Real World Into the Classroom

Why did all of the numbers avoid conversing with Pi at the party? Because he goes on and on forever. –Unknown

Students need to know why they are learning these crucial concepts, as many will think that once school is over, they will never need them again. The fact is—especially with math and numbers—they will need these skills throughout their entire life. Bringing the real world into the classroom helps students make connections with their learning while increasing their engagement. We will start with ideas for bringing the real world into the math classroom followed by bringing more inspiration into the special needs classroom.

Why Math and Special Needs Students Require Real-World Examples

For students, mathematics is a subject that has long been thought of as abstract and disconnected from the real world; of course, we teachers know that this isn't true. Math is just one way of describing the world. However, most curriculums don't teach mathematics this way. Instead, students are often asked to memorize formulas and apply them without any explanation of what the formulas are used for in the real world.

Research shows that connecting math to the real world can have a significant impact on students, particularly those with special needs. This is often a large factor in why students who take physics tend to do better in math; in addition to having extra mathematically based class time, they are seeing the real-world applications of math, and this itself improves student engagement. Many students with special needs struggle with abstract concepts and need concrete examples to understand them. By connecting math to real-world situations, teachers can help students see the practical applications of math and how it relates to their everyday lives.

For example, you might use a cooking lesson to teach fractions or measurement. You could show students how to double or halve a recipe or how to measure ingredients accurately. This approach not only helps students learn math skills but also reinforces life skills that are essential for daily living. By incorporating real-world examples and activities, teachers can create a more dynamic and engaging learning experience for students.

You might use a shopping trip to teach students about money and budgeting or have students create a shopping list, calculate the cost of items, and practice making change, as we discussed in Chapter 2. Making this hypothetical shopping trip about interior decor can even add a geometry-based element, as students will need to calculate areas and diameters to ensure that their chosen furniture fits in a given room. Students with special needs often struggle with motivation, and they need to see a clear connection between what they are learning and their future aspirations. By showing students how math is used in different professions and industries, you can help them see the value of math and how it can help them achieve their goals.

Connecting math to the real world can also help to build students' confidence and self-esteem. Many students with academic support needs struggle with feelings of inadequacy and frustration when it comes to math. By providing them with real-world examples and opportunities to apply math in meaningful ways, you can help students see that they are capable of learning and using math effectively.

Keeping Math Real

By working on real-world projects, students can develop a deeper understanding of mathematical concepts and see how math is used in everyday life. These projects can also be used to tie in other subjects and interests, making math more engaging and accessible. Let's take a closer look at some ideas you can use in the classroom.

Real-World Math Wall

This can be a space where students can post examples of how math is used in the real world. Teachers can encourage students to bring in examples from their daily lives or from news articles, and students can work together to analyze and discuss these examples. This can help students see the relevance of math in their own lives and can be a great way to tie in other subjects, such as science and social studies.

School Water Audit

This project can be used to teach students about the importance of conserving water and can be tied into math concepts such as measurement and data analysis. Students can work together to track water usage in their school over a period of time and analyze the data to identify areas where water is being wasted. They can then develop and implement strategies to reduce water usage and monitor the impact of these strategies over time.

Pretend Restaurant

Students can work together to create a restaurant menu, calculate the cost of ingredients, and determine the pricing for menu items. They can also practice using math skills to calculate tips and taxes. This project can be especially engaging for students with special needs who enjoy hands-on learning and may have an interest in the hospitality industry.

Integrating Math With English and History

Students can analyze historical data—such as population growth or economic trends—and use math concepts to make predictions about the future. They can also use math to analyze literature or historical documents, such as by analyzing the frequency of certain words or patterns in a text.

Math Recipes

Students can work together to create recipes and use math concepts such as measurement, fractions, and ratios to determine the amounts of ingredients needed. They can also use math to scale recipes up or down based on the number of servings needed. This project can be especially engaging for students with special needs who enjoy cooking or have an interest in the culinary arts.

Graphing Halloween Candy

This can be a fun and engaging way to teach math concepts such as data analysis and probability. Students can work together to collect data on the types of candy they receive on Halloween and can use math concepts to analyze and graph the data. They can also use probability to make predictions about the types of candy they are likely to receive based on past data.

Other Ways to Introduce Special Needs Students to Math in the Real World

The news is a fast and free way to bring real-world and relevant topics into the classroom. You can use reports on finances and even sports to incorporate math themes. Look to parents who use math in their careers, and invite them as guest speakers. If you can't find guest speakers, you can have students act out the roles of people in mathematical situations in life. If possible, take a field trip, and use this field trip to reinforce concepts. Finally, give students tangible problems to solve—really, when was the last time you had five apples in one hand and six oranges in the other?

Of course, the types of projects you can use aren't limited to the above; you could even ask your class for some ideas to get them excited and more involved in the process.

In addition to real-world examples of math to inspire students, there is a resource inside the classroom that many teachers don't take advantage of. Considering that children are individuals and learn at their own pace, there will be mini-teachers who will thrive on being given the chance to help their classmates! In the next chapter, we will take a look at peer-based learning and see how you can implement it in your classroom.

Chapter 6:
Peer-Based Learning

A true friend accepts who you are, but also helps you become who you should be. –
Cameron Jenkins & Yaa Bofah

Peer-based teaching occurs when students share concepts, learning with and from other students. It is beneficial because those who have grasped a concept can explain it to others. This type of buddy system improves confidence and well-being. At the same time, it makes excellent use of resources that would otherwise be missed. Instead of having one teacher, the class can suddenly have various teachers.

Why Peer-Based Learning Is a Valuable Strategy

Peer-based learning is where students of similar status (such as age or ability) learn from one another. Studies have shown that 88% of peer-based programs in math had a positive effect on student learning (*Peer Tutoring and Academic Achievement in Mathematics: A Meta-Analysis*, n.d.).

Some students may feel more comfortable learning from a peer and the dynamic that comes from this form of learning. It can change a student's attitude toward learning and increase engagement, learning can be more personal, and there is more cooperation in the classroom. Peer teachers can gain more confidence, and overall, academic results can be higher (Fink, 2020).

For students with special needs, peer-based learning can be especially beneficial. It can improve social skills, as students interact with each other in a supportive and positive environment. Moreover, peer teachers can serve as role models, demonstrating positive behaviors and attitudes toward learning. Students with special needs may also feel more comfortable learning from their peers, as they may be more relatable and better able to understand their unique needs.

How to Prepare Students for Peer-Based Learning

In order to ensure that peer-based learning is effective, it is important to prepare students properly. There are a number of ways that you can do this. Let's take a closer look at some of them:

- **Build a positive classroom culture:** Before implementing peer-based learning, it is important to build a positive classroom culture that fosters respect, cooperation, and empathy. This can be achieved by establishing clear expectations for behavior and encouraging positive interactions among students. You should model and reinforce positive behaviors and provide opportunities for students to practice them.

- **Teach collaboration skills:** Collaboration is a key component of peer-based learning, and students need to be taught how to collaborate effectively. You can do this by teaching collaboration skills such as active listening, constructive feedback, and conflict resolution. These skills can be taught through role-playing exercises and group activities.

- **Establish clear goals and expectations:** It is important to establish clear goals and expectations for peer-based learning. You should communicate these goals and expectations to students and provide them with feedback on their progress. This will help students understand what is expected of them and how they can succeed in the peer-based learning environment.

- **Assign roles and responsibilities:** Assigning roles and responsibilities can help students understand their role in the peer-based learning process. This can include assigning group leaders, timekeepers, note-takers, and other roles. You can rotate these roles to ensure that all students have the opportunity to take on different responsibilities.

- **Provide opportunities for practice:** Students need opportunities to practice peer-based learning before they can do it effectively. You can provide opportunities for practice by incorporating small group activities and projects into their lesson plans. This will give students the chance to work together and practice their collaboration skills in a safe and supportive environment.

- **Provide feedback and support:** You should provide feedback and support to students throughout the peer-based learning process. This can include providing feedback on group projects and individual contributions, as well as offering support and guidance to students who are struggling. You should also monitor group dynamics and step in if there are any issues.

- **Encourage reflection and self-assessment:** Reflection and self-assessment are important components of peer-based learning. You can encourage students to reflect on their experiences and assess their own learning. This can be done through journaling, group discussions, and other reflective activities. By reflecting on their experiences, students can gain a deeper understanding of their own learning and the learning process.

Peer-Based Learning Techniques

There are a couple of methods that can be used for peer-based learning. Each strategy has its own unique benefits and can be used to help students become more comfortable and successful in working with their peers. Let's look at eight of the most common methods:

- **Proctor model:** In this model, students are paired up with a more experienced peer who acts as a guide or mentor. The proctor can help the student with understanding concepts, answering questions, and providing feedback. For example, in a math class, a student struggling with geometry could be paired with a proctor who excels in that subject.

- **Discussion seminars:** Discussion seminars provide a forum for students to share their ideas and perspectives on a particular topic. This can be especially helpful in a math class, where there are often multiple ways to approach a problem. By sharing their approaches and reasoning, students can learn from one another and gain a deeper understanding of the material.

- **Peer support groups:** Peer support groups are small groups of students who meet regularly to discuss their progress and challenges in a particular subject. This can be especially helpful for students who are struggling with math, as they can receive encouragement and support from their peers.

- **Peer assessment schemes:** Peer assessment schemes involve students evaluating one another's work. This can be helpful in a math class, where there are often multiple ways to solve a problem. By evaluating each other's work, students can gain a better understanding of the different approaches and techniques used to solve a problem.

- **Collaborative projects:** Collaborative projects involve students working together to complete a task or project. This can be especially helpful in a math class, where students can work together to solve complex problems or complete a challenging assignment.

- **Cascading groups:** Cascading groups involve students teaching other students who are then responsible for teaching yet more students. For example, in a math class, a student who has mastered a particular concept could teach it to another student, who would then teach it to another.

- **Reciprocal teaching:** Reciprocal teaching involves students taking turns teaching one another. For example, in a math class, one student could explain how to solve a particular problem while the other student asks questions and provides feedback. Then, for the following problem, they would swap roles.

- **Expert jigsaw method:** In the expert jigsaw method, students become experts in a particular topic and then teach that topic to their peers. For example, in a math class, students could be divided into groups, with each group becoming an expert in a particular area of math.

While children will thrive on peer-based learning, it's important not to make too many changes all at once. Some children with social and learning disorders may struggle with adapting to change, and announcing that one student is going to be the teacher could fill them with pride or terrify them. Approach every new technique with gentle curiosity and analyze what works best for each student or class group.

Next, we will look at teaching strategies to improve a student's understanding of numbers and their relationships with a strategy that puts things in perspective before looking at problems.

Chapter 7:
Concrete-to-Visual-to-Abstract

Math—The only place where people buy 64 watermelons and no one wonders why!
–Daniel B. Markham

In this chapter, we will explore what the *concrete, visual, abstract* (CVA) strategy is and how you can implement it in the classroom to help your students make sense of number concepts and simplify algebra expressions and computation.

What Is the Concrete, Visual, Abstract Strategy

The concrete, visual, abstract (CVA) strategy is an approach that helps students understand complex concepts by breaking them down into three stages. The first stage—the concrete stage—involves the use of physical, tangible objects such as coins, shapes, or counters. This stage is important because it allows students to manipulate and interact with objects in a hands-on way, which can help them form a strong foundation of understanding.

The second stage—the visual stage—involves the use of visual representations such as drawings, diagrams, or charts. This stage is important because it helps students connect the concrete objects they have been working with to more abstract ideas. Visuals can make it easier for students to see patterns, relationships, and connections between different concepts, which can help them make sense of more complex ideas.

The final stage—the abstract stage—involves the use of symbols, letters, and numbers to represent concepts. This stage is important because it helps students make the leap from concrete objects and visual representations to more abstract ideas that may not have a physical representation. By this stage, students should have a solid understanding of the concept they are working with and be able to use symbols, letters, and numbers to represent and manipulate it.

You can use the CVA strategy to reinforce learning by going back and forth between stages. For example, you might start with a concrete activity to introduce a concept, move to a visual activity to help students make connections, and then move to an abstract activity to solidify understanding.

How to Develop a CVA Unit

To develop a CVA unit for your classroom, you can follow these steps:

1. Begin by **planning out the abstract steps** needed to solve the problem. For each step, provide clear verbal reasoning that ties it to the solution. This will help you ensure that your approach is sound and logical.

2. **Choose a concrete manipulative** that can accurately represent each step of the problem-solving process. The manipulative should be easy to use and provide a tangible way for students to visualize the steps.

3. Start teaching the concept to **students using the concrete manipulatives** and simpler numbers. This will help them gain accuracy with verbal reasoning by allowing them to physically work through the problem.

4. Once students have demonstrated accuracy with the manipulatives and easier numbers, you can move on to **using visual representations** of the manipulatives with more complex numbers. This will help them build their skills and gain mastery in solving problems with verbal reasoning.

5. Finally, **teach the abstract concept** using increasingly complex examples until students can demonstrate mastery with verbal reasoning. This will help them develop a deep understanding of the problem-solving process and be able to apply it in a variety of contexts.

Examples of CVA in Math

Let's go over some examples of how you can use the CVA method for each of the steps we discussed above:

1. **Plan abstract steps:** Let's say the problem is to teach students how to solve an addition problem. You might plan the abstract steps like this:

    a. Identify the numbers to be added.

    b. Line up the numbers vertically.

    c. Add the ones column.

    d. Carry over any tens to the next column.

    e. Add the tens column.

    f. Continue adding columns until all columns have been added.

2. **Choose concrete manipulative:** For this, you might choose a set of blocks or cubes that can be stacked to represent the numbers being added.

3. **Start teaching with manipulatives and simpler numbers:** For example, start with simple addition problems like 2 + 3 or 5 + 4, using the blocks to physically represent the numbers and help students visualize the addition process.

4. **Use visual representations with more complex numbers:** As students become more comfortable with addition, you can move on to using visual representations like number lines or grids to solve more complex addition problems like 23 + 47 or 65 + 78.

5. **Teach the abstract concept with increasingly complex examples:** Finally, you can teach the abstract concept of addition using more complex examples like multi-digit addition problems with carrying over or addition problems involving decimals or fractions. This will help students develop a deep understanding of the problem-solving process and be able to apply it in a variety of mathematical contexts.

As a teacher, I've always been passionate about finding the most effective teaching approaches to help my students succeed. Over the years, I've tried various methods, but I've found that the CVA approach works best for many of my students.

I love the CVA approach because it is simple yet highly effective. It breaks down complex concepts into smaller, more manageable pieces, which makes it easier for my students to understand. I've found that this approach is particularly helpful for visual learners who need to put things into context before attempting to understand the abstract.

One of the things I appreciate most about the CVA approach is that it is a strategy that can be incorporated into real-world learning. I've used this approach in various circumstances, but I find it especially effective in role-playing scenarios. For instance, if I'm teaching my students about conflict resolution, I'll have them act out different conflict scenarios using concrete manipulatives to represent the different steps of the problem-solving process. This helps them visualize the steps and understand how they relate to the solution. It's been a valuable tool in my teaching toolbox, and I plan to continue incorporating it into my lessons in the future.

Another strategy to help students with math is to become experts in guiding them through word problems. Before this can happen, it's necessary for all students to learn the right language and terminologies that are used in the world beyond the classroom. In the next chapter, we will discuss how you can use the right language in your classroom.

Chapter 8:
Using the Right Math Language

There is a fine line between a numerator and a denominator. Only a fraction of people will find this funny! –Rachel Griffith

While simplifying language may seem like a good idea, it can actually lead to confusion for students who end up having to learn multiple words for the same mathematical concept. In this chapter, we will explore the importance of using the correct math language and provide strategies to help students remember and use this vocabulary effectively.

Why Should You Be Teaching the Correct Math Vocabulary

Mathematics is a language in its own right, and like any language, it has its own unique vocabulary. Understanding and using the correct math vocabulary is necessary for effective communication between students and teachers. When students know the correct math vocabulary, they can communicate their ideas and understanding clearly and precisely. For example, if a student is having trouble with a math problem, they can ask their teacher for help using the appropriate vocabulary. You can then easily understand the student's problem and provide the necessary guidance.

Using the correct math vocabulary also helps your students to understand the subject matter better. When students learn math vocabulary, they are not just memorizing words but also learning the meaning behind those words. By understanding the meaning of the math vocabulary, students can more easily grasp the concepts and ideas that those words represent. For example, when students learn the term "area," they are not just learning a word, they are also learning the concept of the amount of space inside a two-dimensional shape.

Students who understand the correct math vocabulary can more easily tackle complex math problems. When your students are presented with a math problem, they can break it down into smaller parts and use the appropriate math vocabulary to describe each part. This helps them understand the problem better and devise a plan to solve it. For example, when solving a word problem, students can use the math vocabulary they have learned to break the problem down into smaller parts and identify the key information needed to solve the problem.

Using the correct math vocabulary can also lead to improved test scores. When students understand math vocabulary, they are better equipped to answer questions on tests and exams. Tests and exams often require students to use specific math vocabulary to explain their answers. When students use the correct math vocabulary, they are more likely to receive full credit for their answers. Moreover, without an understanding of the correct terminology, even a skillful student may not recognize what they are being asked to do in an exam scenario. Knowing the correct math vocabulary can help your students to avoid confusion and errors on tests and assessments.

Mathematics is used for many real-world applications, from engineering to finance to medicine. Understanding the correct math vocabulary is crucial for success in these fields. By teaching students the correct math vocabulary, you are preparing them for future success in their careers. For example, engineers need to understand the math vocabulary associated with geometry, trigonometry, and calculus to design and build structures.

How Much Vocabulary Is Necessary

It can be difficult to know exactly what vocabulary your students need to learn, especially since for many math teachers the vocabulary is second nature. Below are the most common terms used up to and including high school math and some simple explanations that you can use to teach your students.

Numbers and Operations

These are the foundation of math, and it's essential that students understand the language that goes with numbers and operations. Some key terms include

- **Whole numbers:** These are numbers that are not fractions or decimals, and they include positive numbers (1, 2, 3, etc.) and zero.

- **Fractions:** These represent parts of a whole and are written as a numerator over a denominator—for example, ½ or ¾.

- **Decimals:** These are numbers with a decimal point, such as 0.5 or 3.14.

- **Integers:** These are whole numbers and their negative counterparts (-3, -2, -1, 0, 1, 2, 3, etc.).

- **Rational numbers:** These are numbers that can be expressed as a fraction or decimal.

- **Irrational numbers:** These are numbers that cannot be expressed as a fraction or decimal—for instance, π or $\sqrt{2}$.

Algebra

Algebra is a significant part of high school math, and understanding the vocabulary that goes with it is essential for success. Some key terms include

- **Variables:** These are symbols that represent a value that can change, such as x or y.

- **Equations:** These are statements that show that two expressions are equal—for instance, $2x + 3 = 9$.

- **Inequalities:** These are statements that show that two expressions are not equal—for example, $x + 4 < 7$.

- **Functions:** These are relationships between inputs and outputs, and they can be represented using graphs, equations, or tables.

- **Exponents:** These are numbers that show how many times a base number is multiplied by itself—for example, $2^3 = 2 * 2 * 2 = 8$.

- **Coefficients:** A number placed before a variable that acts as a multiplier for that variable, such as in the case of 2x.

- **Polynomials:** These are expressions that contain variables and coefficients, and they can be added, subtracted, multiplied, and divided.

Geometry

Geometry involves the study of shapes, sizes, and positions of objects. Some key terms include

- **Points, lines, and planes:** These are the building blocks of geometry and are used to describe shapes and angles.

- **Angles:** These are measures of the space between two lines that meet at a point.

- **Triangles:** These are polygons with three sides, and they can be classified by their angles—acute, right, or obtuse—or their sides—scalene, isosceles, or equilateral.

- **Quadrilaterals:** These are polygons with four sides, and they can be classified by their angles—rectangle, square, rhombus, parallelogram—or their sides—trapezoid.

- **Circles:** These are shapes that are defined by an infinite set of points that are equidistant from a center point.

- **Volume and surface area:** These are measurements of the space inside or outside of a 3D object.

Statistics and Probability

Statistics and probability involve the study of data and the likelihood of events. Some key terms include

- **Mean, median, and mode:** These are measures of central tendency that are used to describe a set of data.

- **Range:** This is the difference between the highest and lowest values in a set of data.

- **Probability:** This is the likelihood of an event occurring.

Ratio and Proportion

Ratios are a way of comparing two or more quantities of the same unit. They are expressed in the form of a:b, where a and b are numbers. Proportion is the equality of two ratios. If a:b=c:d, then a, b, c, and d are said to be in proportion. This concept is commonly used in geometry and algebra and is essential for solving problems related to scaling and similar figures.

Exponents and Radicals

Exponents are a shorthand notation for repeated multiplication. For example, 3^4 means 3 multiplied by itself four times, or 3 * 3 * 3 * 3 = 81. Radicals, on the other hand, are the inverse of exponents. They represent the nth root of a number, where n is the index of the radical. For example, the square root of 25 is represented by the symbol $\sqrt{25}$ and is equal to 5. Exponents and radicals are used extensively in algebra and calculus and are also important in geometry and trigonometry.

Sequences and Series

A sequence is a list of numbers that follows a certain pattern or rule. A series is the sum of the terms in a sequence. Arithmetic sequences have a constant difference between consecutive terms, while geometric sequences have a constant ratio between consecutive terms. Series can be finite—i.e. the sum of a specific number of terms—or infinite—the sum of an infinite number of terms. Sequences and series are important in algebra and calculus and are also used in probability and statistics.

Domain and Range

The domain of a function is the set of all possible input values, while the range is the set of all possible output values. The domain and range can be represented as intervals on a number line or as sets of numbers.

Coordinate Plane

The coordinate plane is a two-dimensional plane that is used to plot points and graph functions. The *x-axis* represents the horizontal axis, while the *y-axis* is vertical.

Slope

Slope is a measure of the steepness of a line. It is represented as a ratio of the change in y divided by the change in x and is given by the formula: slope = $(y_2 - y_1) / (x_2 - x_1)$.

Linear Regression

Linear regression is a statistical method used to find the best-fit line that describes the relationship between two variables. It is commonly used to analyze data and make predictions.

These are just some of the important vocabulary terms that high school math students need to know in order to be successful in their classes. It is essential that you emphasize the importance of learning and understanding these terms, as they form the foundation of math education. You might want to consider investing in a mathematical dictionary if you find that your students are lacking in an understanding of a lot of vocabulary.

Frayer Model

The Frayer Model is a graphic organizer designed to assist students in building their vocabulary and conceptual understanding of math. It consists of four quadrants wherein they can write the definition, examples, non-examples, and characteristics for each math term.

Teachers can utilize the Frayer Model in math by providing students with a list of key terms throughout the unit or lesson. With each word, students fill in quadrants on the Frayer Model accordingly.

Students can write the mathematical definition of a term in the definition quadrant. Then, in the examples quadrant, they provide examples of its usage in math problems or scenarios, while non-examples clarify misconceptions. Finally, in characteristics, the student should provide any additional information necessary to define or distinguish this term from related terms.

The Frayer Model can aid students in developing an intimate understanding of math vocabulary and concepts, serving as a helpful reference throughout the unit or lesson.

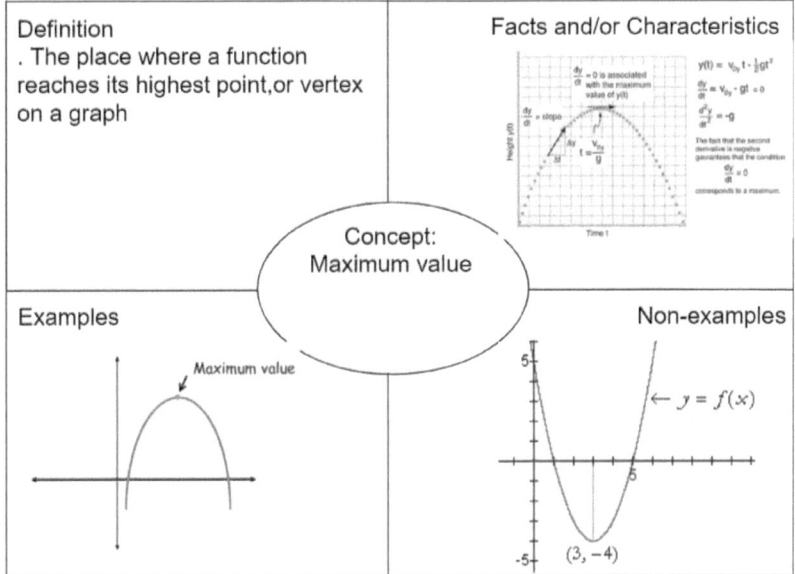

Understanding Word Problems

Fisher and Frey (2014) propose six steps for deciphering a complex text:

1. **Preview the text:** Before reading a text, take some time to preview it. Take note of the title, headings, subheadings, and any graphics or illustrations to get an overview of what the text will cover and prepare yourself to read it more comfortably.

2. **Activate prior knowledge:** Consider what you already know about the topic of the text to help you make connections and gain a deeper understanding.

3. **Chunk the text:** Break up the text into smaller chunks, and read each section one at a time. At the end of each section, take a moment to summarize your understanding so that all information is understood.

4. **Read the text carefully and slowly:** Read the text slowly and carefully, paying particular attention to unfamiliar words or phrases. Utilize context clues and prior knowledge to decipher their meaning.

5. **Ask questions:** As you read, ask yourself some questions. What is the author trying to convey? Why did they create this text? What evidence supports their assertions?

6. **Summarize and reflect:** After reading the text, synthesize what you have learned and consider what more work remains. Doing this helps you retain information and prepare for further reading or discussion.

Reading a math word problem requires specific steps that differ from reading complex texts, however. So, here are five tips to help your students efficiently read and solve math word problems:

1. **Be careful when identifying the problem:** Begin by reading through the problem carefully and identifying its objectives. Pay special attention to any language used or specific numbers or units mentioned.

2. **Identify <u>known</u> and <u>unknown</u> quantities:** Recognize which quantities are known and which remain uncertain. Doing this will enable you to decide what needs solving and what information is available.

3. **Calculate the mathematical operation(s) needed:** Determine which mathematical operations are necessary to solve the problem. For example, these operations might involve addition, subtraction, multiplication, division, or some combination thereof.

4. **Solve the problem:** Use the identified mathematical operations to solve the puzzle. Then, work through it step-by-step, keeping track of any units or conversions that may be needed.

5. **Verify your answer:** After solving the problem, double-check your solution to ensure it makes sense in the puzzle context. Confirm that you have answered all questions and double-check all calculations for accuracy.

By following these steps, any student can master math word problems and develop their problem-solving capabilities in math (Fisher & Frey, 2015).

During the last seven years, while working with students who struggle with math, my students have developed the following additional strategy for solving a word problem:

1. Read the word problem two times—silently the first time, but out loud the second.

2. Number every line of text for a discussion of the problem.

3. Highlight or underline words essential to understanding the context of the problem, especially any words you do not understand.

4. Circle all numbers, including those words which are numbers.

SBAC Construct-Relevant Vocabulary for Mathematics

The Smarter Balanced Assessment Consortium (SBAC) is a consortium of states working together to develop and implement a standardized test that measures student achievement in English language arts (ELA) and mathematics. The SBAC assessment is used by the state of California for federal accountability purposes, and students in grades 3 through 8 and grade 11 take this test in the spring of each school year.

One important aspect of preparing for the SBAC assessment is for students to learn the content vocabulary that is part of the annual assessments. Vocabulary instruction should be part of the curriculum throughout the school year, so students have a deep understanding of the terminology they will encounter on the test.

The SBAC assessment measures student achievement in a range of mathematical concepts, including algebra, geometry, and statistics. To prepare students for the assessment, it is essential to provide instruction on "construct-relevant vocabulary," which refers to any mathematics term that students should know because it is essential to constructing the content area.

Mathematics vocabulary can be challenging for students, and therefore, educators need to make sure they understand and remember the words. You must use explicit instruction to ensure that students understand the meaning of mathematical terms and how they relate to the subject matter. Students need to learn how to use math vocabulary in context so that they can apply the terms when solving problems.

To help educators prepare for the SBAC assessment, the Consortium provides a comprehensive list of vocabulary terms that students should know. The SBAC mathematics vocabulary is organized by grade level and includes terms such as "area," "bar graph," "diameter," and "integer." By using this list, you can ensure that you are covering all the essential terms that students need to know.

The importance of mathematics vocabulary instruction goes beyond just preparing students for the SBAC assessment. Understanding math vocabulary is essential for success in math and other subjects that use mathematical concepts. Students who struggle with math vocabulary may have difficulty understanding word problems or communicating their thinking effectively.

To help students develop a deep understanding of math vocabulary, it is recommended that educators should use a variety of instructional strategies. Explicit instruction is crucial, but students also need opportunities to use math vocabulary in context. This can be done through word problems, math discussions, and other activities that require students to apply mathematical terms to real-world situations.

You can also incorporate technology into vocabulary instruction to make it more engaging and interactive. Online math games and quizzes can help students practice and reinforce their understanding of math terms, while interactive whiteboards can be used to visually represent mathematical concepts and vocabulary.

Mathematics vocabulary instruction is an ongoing process. Vocabulary should be reviewed and reinforced throughout the school year, not just in the weeks leading up to the SBAC assessment. You can incorporate math vocabulary into daily instruction and review key terms regularly to ensure that students are mastering the terminology.

Naturally Adding Vocabulary to Lessons

Having a math word wall is a great way to promote math vocabulary acquisition and application in the classroom. To make it even more effective, you should ensure that the math word wall only displays math vocabulary that is relevant to the current unit being taught. This helps students to focus on the most important and useful vocabulary for that unit.

Using visual cues with images can be really helpful for visual learners, as it helps to connect the vocabulary word with a visual representation. For example, a picture of a parallelogram can be posted next to the word "parallelogram" on the math word wall. Another strategy is to have students keep a vocabulary book where they draw and define each word they learn. This can be helpful for artistic learners who may enjoy the creative aspect of drawing the word.

For kinesthetic learners, encouraging students to use gestures to remind them of key terms can be effective. For example, a student might make a hand gesture in the shape of a right angle when reminded of the phrase "right angle." This can help them to remember and retain the vocabulary.

Playing math charades is a fun and engaging activity that can incorporate gestures and movements while also reinforcing math vocabulary. In this game, students act out a math vocabulary word and other students guess the word. This can be done in small groups or as a whole-class activity.

Giving credit to students when they find or use a vocabulary word in the real world can also be motivating and reinforce the importance of learning math vocabulary. For example, a student might see the word "quadrilateral" on a sign while out in the community and point it out to the teacher. The teacher can then give the student credit for recognizing and using the word in context.

Using fill-in-the-blank exercises can also help reinforce vocabulary acquisition and application. For example, a sentence might be provided with a missing vocabulary word that students need to fill in. This can be done as a homework assignment or in class as a review activity.

It's important to use the correct vocabulary yourself, but include student-friendly definitions. This helps students to understand the word without feeling overwhelmed by technical jargon. Giving students multiple exposures to the word in various contexts can also help them internalize the vocabulary word. By using a combination of these strategies, you can help your students with learning disabilities and diverse learning styles acquire and apply math vocabulary in the best way possible.

I remember a time back when I was in school and my math teacher gave us a list of 30 words to learn over the weekend for a test on Monday. I remember feeling overwhelmed and frustrated with the task at hand. It seemed like such a daunting and boring task that I didn't even know where to start.

Looking back now, I realize that this approach to teaching math vocabulary was not effective. It lacked creativity and failed to engage students in a meaningful way. Fortunately, those days of teaching vocabulary through brute force rote memorization are over. As a math teacher, I am motivated to find unique and fun ways for my students to learn math vocabulary.

I believe that learning math vocabulary should be an enjoyable and interactive experience for students. That's why I have implemented various strategies in my classroom to make it more engaging. For example, I have a math word wall that is specific to the unit we are currently working on. The wall includes pictures and visual cues to help students make connections with the vocabulary words. I also incorporate games like math charades, which encourages students to act out math vocabulary words while their classmates guess what they are trying to convey. This not only helps students remember the words but also allows them to have fun while doing it.

I encourage students to find and use math vocabulary words in real-world situations. This promotes a deeper understanding of the vocabulary and its applications. When a student uses a vocabulary word in class, I make sure to give them credit and praise for their effort. By using these strategies, I am able to engage with my students and make math vocabulary learning an enjoyable experience.

The schema-based instruction requires correct language and following a set of directions in order to encourage children to solve word problems that can seem overwhelming at first. Once students learn the process, they will be able to become more independent learners. In the next chapter, we will go over how you can use this method to teach your students.

Chapter 9:
Schema-Based Instruction

Math Problem: John has 32 candy bars. He eats 28. What does he have now? Diabetes! –Unknown

Many standardized tests require students to solve math word problems, which can be confusing as they often involve complex vocabulary. However, students can use a schema-based strategy to simplify the process. This approach involves four steps: identifying the problem, setting it up, solving it, and checking the solution. By following these steps, students can tackle math word problems with confidence and accuracy. In this chapter, we will explore the schema-based strategy in depth, providing a concise and effective tool for problem-solving.

Compare Word Problems

Compare problems involve the comparison of two separate sets and emphasize the relationship between them. This type of problem has also been referred to as a "difference" problem. The three pieces of information in a compare problem are the compared, referent, and difference sets.

Bigger # (B) − Smaller # (S) = Difference (D)

How Schema-Based Instruction Came About

The schema-based instruction method is a teaching strategy that emphasizes the semantic structure of a problem as well as the mathematical structure. In the 1980s and '90s, psychological and mathematical educational researchers began investigating the cognitive issues involved in problem-solving. They noticed that many students struggled with solving math word problems not because they lacked the mathematical skills but because they had difficulty understanding the problem itself.

Further research revealed that students often approached problem-solving in a haphazard way, without a systematic approach. They would often skip important steps and fail to properly identify the type of problem they were trying to solve. This lack of organization and structure led to confusion and errors.

To address this issue, researchers proposed the schema-based instruction method, which involves teaching students a set of steps for identifying and solving problems. The method emphasizes the importance of understanding the semantic structure of a problem or the underlying meaning of the problem, as well as the mathematical structure.

By breaking down problems into manageable steps, students can more easily identify the relevant information, set up the problem correctly, and solve it accurately. The schema-based instruction method has been found to be an effective approach to teaching problem-solving skills, as it provides students with a clear framework for approaching math word problems.

Schemas for Different Word Problem Structures

A schema is a mental framework or structure that helps individuals organize and interpret information. In the context of word problems, schemas are problem-solving strategies in which your students identify the type of problem they are dealing with and use a systematic approach to solve it. Different types of word problems have different structures, and each structure requires a different schema to solve it effectively.

Let's take a look at some common word problem structures and the schemas that can be used to solve them.

Addition and Subtraction

Addition and subtraction word problems are among the most common. They involve adding or subtracting two or more numbers to find a solution. The schema for solving these types of problems involves four steps:

1. Read the problem carefully, and identify the numbers involved.

2. Determine whether the problem requires addition or subtraction.

3. Add or subtract the numbers accordingly.

4. Check your answer to ensure accuracy.

For example, consider the following problem: Samantha has 5 apples. She buys 3 more apples. How many apples does she have in total?

To solve this problem, we would use the addition schema. We would first identify the numbers involved (5 and 3), determine that addition is required, add the numbers (5 + 3 = 8), and then check our answer.

Multiplication and Division

Multiplication and division word problems are another common type of word problem. They involve multiplying or dividing two or more numbers to find a solution. The schema for solving these types of problems involves five steps:

1. Read the problem carefully, and identify the numbers involved.

2. Determine whether the problem requires multiplication or division.

3. Set up the problem using the appropriate operation.

4. Solve the problem by multiplying or dividing the numbers.

5. Check your answer to ensure accuracy.

For example, consider the following problem: John has 3 bags of candy, with 8 pieces of candy in each bag. How many pieces of candy does John have in total?

To solve this problem, we would use the multiplication schema. We would first identify the numbers involved (3 and 8), determine that multiplication is required, set up the problem ($3 * 8 = ?$), multiply the numbers ($3 * 8 = 24$), and then check our answer.

Ratio and Proportion

Ratio and proportion word problems involve comparing two or more quantities in relation to one another. They require the use of ratios and proportions to solve. The schema for solving these types of problems involves six steps:

1. Read the problem carefully, and identify the quantities involved.

2. Determine the relationship between the quantities—for example, direct proportion or inverse proportion.

3. Write the ratio or proportion that represents the relationship between the quantities.

4. Solve the ratio or proportion using cross-multiplication.

5. Simplify the answer, if necessary.

6. Check your answer to ensure accuracy.

For example, consider the following problem: A recipe calls for 2 cups of flour for every 3 cups of sugar. If the recipe requires 9 cups of sugar, how many cups of flour are needed?

To solve this problem, we would use the ratio and proportion schema. We would first identify the quantities involved (flour and sugar), determine the relationship between the quantities (direct proportion), write the proportion ($2/3 = x/9$), solve the ratio using cross-multiplication ($3x = 18$), simplify the answer ($x = 6$), and then check our answer.

Putting Schema-Based Instruction Into Action

So, now that you've seen some examples of schemas, let's take a look at how you can start using them in the classroom.

1. **Introduce Schemas:** The first step in implementing schema-based instruction is to introduce students to the concept of schemas. This can be done by explaining what a schema is and how it can help them solve math problems more efficiently. It is also important to provide examples of different types of schemas that will be used in the classroom.

2. **Identify Problem Types:** The next step is to help students identify different types of math problems. This can be done by providing examples of each problem type and explaining the characteristics that make them unique. For example, addition and subtraction problems involve adding or subtracting two or more numbers, while ratio and proportion problems involve comparing two or more quantities in relation to one another.

3. **Introduce Schemas for Each Problem Type:** Once students understand the different types of math problems, it is time to introduce the corresponding schemas for each problem type. This can be done by modeling the use of the schema and guiding students through the process of solving problems using the schema.

4. **Practice with Guided Examples:** After introducing the schemas, it is important to provide students with guided practice opportunities. This can be done by giving examples of problems and guiding students through the process of using the corresponding schema to solve them. During this stage, teachers should provide feedback and support to help students master the problem-solving process.

5. **Independent Practice:** Once students have had ample guided practice, it is time to provide independent practice opportunities. This can be done by providing students with a set of problems and asking them to solve them independently using the appropriate schema. During this stage, teachers should monitor student progress and provide feedback as needed.

6. **Assessment:** Assessment is a critical component of schema-based instruction. It is important to assess student understanding of the schemas and their ability to apply them to different problem types. This can be done through a variety of assessments such as quizzes, tests, or performance tasks.

Tips for Successful Implementation

Implementing schema-based instruction in the classroom can be challenging, but with some tips and strategies, it can be a success:

1. **Start small:** Introduce one schema at a time, and gradually build on it as students become more comfortable with the process.

2. **Use visuals:** Visual aids, such as diagrams or graphic organizers, can help students understand the problem-solving process and remember the steps involved.

3. **Provide feedback:** Feedback is essential for student learning. Provide students with feedback on their problem-solving process, and encourage them to reflect on their work.

4. **Use real-world examples:** Use real-world examples that are relevant to students to help them see the practical application of problem-solving strategies.

5. **Differentiate instruction:** Differentiate instruction to meet the needs of all learners. Provide additional support for struggling students, and challenge advanced learners with more complex problems.

Remember to reflect on past plans and identify areas where schema-based instruction could have been used to enhance student learning. By analyzing recent plans, you can determine which lessons could have been more effective if a schema-based approach had been utilized. Additionally, by looking ahead to upcoming plans, you can adapt and incorporate a more structured approach to better support student understanding and retention of key concepts.

Building on effective teaching strategies, in the next chapter, we will delve into how you can support students in retaining their learning through a variety of techniques tailored to meet the needs of diverse learners. By utilizing these strategies, you can help ensure that all students, regardless of learning style or ability, are able to engage with and retain the material covered in class.

Chapter 10:
Retention Techniques

What do you get when you cross a computer with an elephant? Lots of memory! –
Daventry Library

This chapter will explore alternative methods for studying and retaining information, including mind mapping, the Leitner System, and the Feynman Technique. By introducing students to these diverse techniques and systems, teachers can empower them to take ownership of their learning and choose study methods that work best for them. Through encouraging independence and confidence in their abilities, students can be empowered to take an active role in their education, leading to more meaningful and effective learning outcomes. Many individuals believe that they either have a good or bad memory and that this trait is fixed and unchangeable. However, with a growth mindset, people can improve their memories and develop effective learning strategies.

As teachers, it is essential to recognize that not all students learn in the same way and that traditional study methods may not be effective for everyone. While teachers often encourage students to study, they rarely provide diverse and effective methods to retain information. In this chapter, we will explore the importance of a growth mindset and discuss various learning strategies that can help students improve their memories and retain information effectively.

The PQ4R Method

As students progress through high school, they are often faced with an increasing amount of material to learn and retain. Whether it be for math, reading, or other subjects, effective study strategies are extremely important. One such strategy is the PQ4R method, an acronym that stands for preview, question, read, reflect, recite, and review. While it is more commonly used for reading, this method can also be applied to high school math, particularly for word problems or reviewing chapters.

The PQ4R method is a comprehensive approach to learning that involves active engagement with the material. By following the six steps outlined below, your students can better organize information, improve comprehension, and retain information.

1. **Preview:** The first step of the PQ4R method involves previewing the material. This step is crucial as it helps students understand what to expect from the text and what is most important. Students should begin by scanning the pages and looking for headings, subheadings, and important vocabulary words. This step can help them get a sense of what the text is about and what key concepts will be covered.

2. **Question:** Once the material has been previewed, students should then generate questions about what they expect to learn. This step helps students to focus on the material and create a sense of purpose for their reading. Questions can be as simple as, "What is this chapter about?" or "What do I already know about this topic?" By generating questions, students can begin to actively engage with the material and set goals for their learning.

3. **Read:** After generating questions, students should then begin to read the text. This step involves carefully reading, taking notes, and highlighting important information. Students should pay attention to the key concepts, definitions, and examples provided in the text. They should also make connections to their previous knowledge and try to visualize the material to help it make sense.

4. **Reflect:** Once the reading is complete, students should take a moment to reflect on what they have learned. This step involves thinking about the key concepts involved and trying to make connections between the material and the real world. It is also an opportunity to clarify any misunderstandings or areas of confusion. Reflecting on the material can help students to better understand the material and to retain the information more effectively.

5. **Recite:** The recite step of the PQ4R method involves actively recalling the material. This step can be done by writing down key points, listing the steps of a problem-solving schema, or talking about the material with someone else. The act of recalling the material helps to solidify the information in the student's memory and makes it easier to retrieve later.

6. **Review:** The final step of the PQ4R method is to review the material. This step involves considering the main points of the material and identifying which ones are still unclear. Students should take the time to go back through their notes, review key concepts, and ask questions if needed. Reviewing the material helps to reinforce the learning and makes it easier to recall later.

Retrieval Practice

Retrieval practice is a proven effective way to help students retain and recall information in the long term. Rather than simply reading or taking notes, retrieval practice involves actively recalling information that is already stored in the brain. This method of learning is based on the idea that the act of recalling information helps to strengthen memory, making it easier to remember that information in the future.

A good way to implement this is through practice tests. This can involve creating a set of questions based on the material that has been covered in class and then having students complete the questions on their own. This could be in the form of multiple choice questions, short answer questions, essay questions, or even classroom games.

You can also use question-and-answer sessions. In these sessions, you ask students questions based on the material that has been covered in class and then encourage students to respond with their own answers. This method can be particularly effective when used in group settings, as it allows students to share their own perspectives and learn from one another.

Consider starting each class by asking students what they remember from the previous class. This encourages students to actively recall information that they have already learned, helping to reinforce that information in their memory. It also allows you to see what information your students are remembering and what information you might need to revise.

You should also provide students with frequent opportunities to practice recalling information. This can involve incorporating practice tests and question-and-answer sessions into regular classroom activities, as well as assigning homework that requires students to recall information they have learned in class—or practice their research skills if they weren't listening.

Retrieval practice can also help students develop stronger critical thinking and problem-solving skills. By actively recalling information and applying it to new situations, students are able to develop a deeper understanding of the material and are better equipped to apply that knowledge in real-world settings.

Spaced Practice

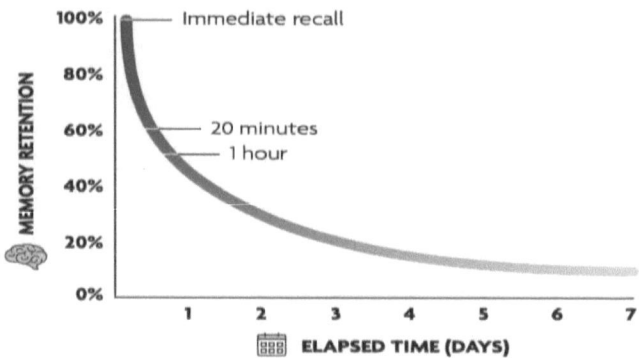

This curve demonstrates how information students learn is lost over time when there is no attempt to consciously review & remember the learned material

Spaced practice is a highly effective technique that involves spreading out learning over an extended period of time. This method is different from the traditional method of learning wherein students cram for exams or learn all the content in one week. With spaced practice, students can improve their retention of material by taking a "little and often" approach, which allows them to avoid the pitfalls of overloading their working memory.

The idea behind spaced practice is based on the concept of the spacing effect, which suggests that information is more effectively learned when it is studied in spaced intervals rather than all at once. This means that instead of studying for several hours in one sitting, students can break up their study sessions into shorter, more frequent intervals over a more extended period of time. For example, instead of studying for three hours in one day, students can be more effective if they study for an hour each day for three days.

To implement spaced practice, you can introduce the concept to your students and explain how it works. You can then provide students with a schedule or planner to help them organize their study sessions. You can also recommend a specific amount of time for each study session, depending on the material that needs to be learned.

For example, if students are studying for a math exam, they can be given a practice problem set that they can work on over a period of a few days. Each day, they can complete a few problems and review the material from the previous day. This way, they will be able to maintain their knowledge of the material and reinforce their learning.

To ensure that students are retaining the information, consider using formative assessments—such as quizzes or class discussions—to gauge their understanding of the material. This can help identify any areas where your students may be struggling and provide an opportunity to review and reinforce the material.

You should also encourage students to use spaced practice outside of the classroom. For example, students can set up study groups or study buddies with whom to review material and practice retrieval of information. This can help them to stay on track and motivated.

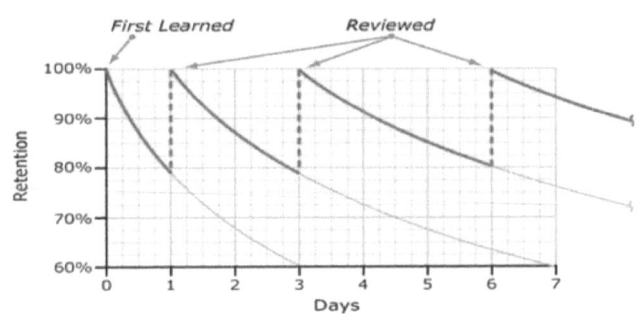

Typical Forgetting Curve for Newly Learned Information

The Feynman Technique

The Feynman technique is a learning strategy that can help your students better understand complex concepts. It involves breaking down a concept into simple terms and explaining it in a way that anyone can understand. This technique is named after Richard Feynman, a Nobel Prize-winning physicist who was known for his ability to explain complex scientific concepts in simple terms.

To use the Feynman technique in the classroom, follow these steps:

1. **Choose a concept:** The first step is to choose a concept or topic that needs to be understood. This could be a challenging math problem, a difficult scientific concept, or a complex historical event.

2. **Learn the concept:** You should take the time to learn the concept thoroughly, making sure you understand all the key components and how they relate to each other.

3. **Explain the concept to a 12-year-old:** Once you have a good understanding of the concept, try to explain it in simple terms that a 12-year-old (or younger) can understand. This means avoiding jargon, technical terms, and complex explanations. When dealing with students of any age, ensure that you know how to phrase conceptual information in such a way that someone five years younger than them could understand it.

4. **Reflect on the explanation:** After explaining the concept, reflect on your explanation and identify any areas that were unclear or confusing. Revise your explanation to make it even simpler and more straightforward if needed.

5. **Combine with peer-based learning:** The Feynman technique can also be combined with peer-based learning. Students can work in pairs or small groups, taking turns to explain concepts to each other using their own simple language and explanations, informed by the information you have given them.

This technique can be used across a wide range of subjects and topics, not just math. It is especially useful for students who struggle to understand complex concepts or who have difficulty with abstract ideas.

The Feynman technique is a powerful tool for improving learning outcomes, as it encourages students to engage deeply with the material, identify areas of confusion, and develop clear and simple explanations. By using this technique, you can help your students to develop a deeper understanding of complex concepts and improve their ability to retain and apply information.

The Pomodoro Technique

The Pomodoro Technique is a time management method that can be particularly helpful for neurodivergent students, including those with ADHD and ASD. This technique can help students with focus, single tasks, time management, reducing stress, and increasing productivity.

To implement the Pomodoro Technique, you can follow these steps:

1. **Set up a timer:** The Pomodoro Technique involves working for a set amount of time and then taking a break. The typical work period is 25 minutes, with a 5-minute break afterward. To make it easy for students to keep track of time, set up an easily visible timer that counts down the 25 minutes and then starts the 5-minute break automatically. There are many apps and websites available for this purpose, or you can use a physical timer.

2. **Explain the technique:** Explain to the students how the Pomodoro Technique works and why it can be helpful. Make sure they understand that they will work for 25 minutes, take a 5-minute break, and then repeat this cycle until they have completed four Pomodoros. After four Pomodoros, they can take a longer break of 20–30 minutes. Encourage them to use this time to rest, recharge, or do something enjoyable.

3. **Provide visual aids:** As mentioned above, it's important to provide visual aids that students can see. A stopwatch bomb is a fun visual aid that can help students keep track of time. You can also use a whiteboard or a poster to display the current Pomodoro cycle, with check marks or stickers for each completed Pomodoro.

4. **Set clear goals:** Encourage students to set clear goals for each Pomodoro cycle. For example, they might aim to complete a certain amount of homework, read a chapter of a book, or practice a specific skill. This can help them stay focused and motivated during the work period.

5. **Adjust as needed:** Remember that every student is different, and some may need more or less time for each Pomodoro cycle. Encourage students to experiment with the timing and adjust it as needed to find what works best for them. They may also find it helpful to adjust the length of the breaks or to take breaks more or less frequently.

By implementing the Pomodoro Technique, you can help neurodivergent students stay focused, manage their time more effectively, and reduce stress. With consistent practice, they can develop the habit of working in focused bursts and taking regular breaks, leading to improved productivity and well-being.

As a teacher, I often feel overwhelmed by the amount of material that I need to cover in a short period of time. I know that my students need to retain as much information as possible, but sometimes, it seems like an impossible task. However, I've come to realize that small changes can make a big difference.

One of the changes that I made was to start using flashcards more regularly. At first, I was hesitant because it seemed like a lot of work to prepare them. But then I had an idea: Why not involve the students in making them? This would not only save me time, but it would also help the students to retain the information better. So, I asked my students to work in pairs and create flashcards for the key concepts that we were studying. They were excited to take on this task, and I was pleased to see them working collaboratively. Once they were finished, I reviewed the flashcards and made any necessary corrections. The result was amazing! The students loved using the flashcards, and I noticed that they were retaining the information better. They were also more engaged in class discussions and seemed more confident when it came to answering questions. I was thrilled to see that my efforts had paid off.

As the years went by, I continued to use the same set of flashcards. I would make minor changes each year, but for the most part, they remained the same. It was a relief to not have to create new materials each year, and the students were happy to use the familiar flashcards. Looking back, I realized that the hard work and effort preparing hundreds of flashcards only made us more determined to use them. The students loved them, and I noticed a difference, which made it all worthwhile. It's true that small changes can prevent teachers and students from becoming overwhelmed, and in this case, it led to better learning outcomes for everyone involved.

Our final strategy and chapter aims to consolidate all the information and insights gained so far and help you to create a comprehensive plan to develop optimal math skills for special needs students.

Chapter 11:
Creating Your Plans and Activities

This final quote is one to remind teachers that although a plan is essential, it's not what makes a great teacher!

Everyone who remembers his own education remembers teachers, not methods and techniques. The teacher is the heart of the educational system. –Sidney Hook

Although teachers must adhere to a prescribed curriculum, they still have the autonomy to choose their teaching methods. This chapter aims to offer practical guidance and techniques to assist you in effectively planning lessons for an entire term, incorporating stimulating activities that can be tailored to suit diverse learners.

A Step-By-Step Guide to Creating Special Education Lesson Plans

Creating special education lesson plans can be a daunting task, but by following these step-by-step guidelines, you can effectively plan and implement lessons that meet the unique needs of your students.

Start With the Learning Target

Starting with the learning target is an important step in creating effective lesson plans for special needs students. By identifying specific learning objectives and clearly defining the learning target, you can ensure that your lessons are targeted toward the needs of your students and aligned with their individualized education program (IEP) goals.

One of the challenges of teaching students with IEPs is that they often have a diverse range of needs and abilities. It can be difficult to create lesson plans that meet the needs of all students, especially when they are working at different levels. However, by starting with the learning target, you can create lessons that are focused and targeted toward specific goals, making it easier to differentiate instruction and meet the needs of individual students.

When identifying the learning target you should examine the specific skills and knowledge that your students need to master. This can involve breaking down larger concepts into smaller, more manageable objectives that can be taught and assessed. For example, if the overall concept is to learn how to use algebra, the learning target might be to identify the components of an algebraic equation or learn how to use letters as placeholders for numbers.

Once the learning target has been identified, it's important to clearly define it so that both you and the student understand what they are working towards. The learning target should be written in a way that is measurable and specific so that progress can be easily tracked and assessed. For example, a clear learning target might be, "By the end of the lesson, students will be able to use algebra to solve business problems." This clearly defines what the students will be able to do at the end of the lesson, making it easier to assess whether they have met the learning target.

You also need to ensure that the learning target aligns with each student's IEP goals. The IEP is a legal document that outlines the specific goals and objectives of each special education student. By aligning the learning target with the IEP goals, you can ensure that their lesson plans are tailored to the specific needs of each student. This helps to ensure that students are making progress toward their individual goals and that they are receiving the support and accommodations that they need to be successful.

The learning target should also be aligned with state and national academic standards. These standards provide a framework for what students should know and be able to do at each grade level. By aligning the learning target with these standards, you can ensure that their lessons are meeting the necessary criteria for academic achievement.

When creating lesson plans for special education students, you should keep in mind that the learning target may need to be modified or adjusted based on the needs of individual students. Some students may require more support or scaffolding to achieve the learning target, while others may be able to work towards more advanced objectives. You should be prepared to adjust lesson plans as needed and ensure that everyone is making progress toward their goals. This target will be the focus of the lesson and will guide all other aspects of planning, including assessment, instruction, and differentiation. Here are some detailed instructions on how to effectively start with the learning target:

1. **Identify the academic or skill-based goal:** Begin by considering what specific academic or skill-based goal you want your students to achieve. This goal should align with state or national standards and should be appropriate for your students' age and grade level.

2. **Write the learning target:** Once you have identified the academic or skill-based goal, write a clear and concise statement that defines the specific learning target. This statement should describe what students will be able to do or know as a result of the lesson. The learning target should be measurable and observable so that you can assess whether or not students have achieved it. It should also be realistic in the timespan available. For instance, learning targets for each class will be vastly different from the overall goals of each week, unit, or semester of learning.

3. **Share the learning target with your students:** Before starting the lesson, share the learning target with your students. This will help them to understand what they will be learning and why it is important. Additionally, sharing the learning target can help to increase student motivation and engagement.

4. **Align the learning target with student needs:** Consider the unique needs of your students and how the learning target can support their learning and growth. For example, if you have students with disabilities, you may need to modify the learning target or provide additional accommodations to help them achieve it.

5. **Consider different levels of achievement:** Not all students will achieve the learning target to the same level or in the same way. Consider how you can differentiate instruction and assessment

to meet the needs of all learners. This may include providing additional support for struggling students or challenging activities for high-achieving students. It might also involve using qualitative feedback on individual student improvement rather than grading every pupil according to the same standard.

6. **Assess student progress towards the learning target:** Throughout the lesson, monitor and assess student progress towards the learning target. Use formative assessments—such as quick checks or exit tickets—to gather data on student understanding and adjust instruction as needed.

Keep Prerequisite Skills in Mind

Prerequisite skills are the foundational skills that students need in order to achieve the learning target. By identifying these skills and ensuring that the lesson builds upon them, you can provide targeted support and remediation where necessary, helping to ensure that all students are able to make progress toward their goals.

Identifying prerequisite skills requires a careful analysis of the learning target and an understanding of the specific skills and knowledge required to achieve it. If the learning target is to solve a complex mathematical problem, prerequisite skills might include understanding the fundamental concepts and principles of mathematics, being able to perform basic arithmetic operations, and understanding how to apply mathematical formulas and equations to solve problems. By identifying these skills, you can ensure that lessons are focused on building towards the larger goal of solving increasingly complex problems or building the algebraic skills that are themselves prerequisites for the introduction of calculus.

Once prerequisite skills have been identified, you should ensure that the lesson builds on these foundational skills. This might involve providing explicit instruction and practice on the prerequisite skills before moving on to the larger learning. For example, if a prerequisite skill is being able to solve complex mathematical problems, the teacher might provide explicit instruction on relevant mathematical concepts, equations, and formulas and provide practice activities to reinforce these skills.

Remember to provide support and remediation where necessary. Special education students often require additional support and scaffolding to master prerequisite skills and achieve the learning target. This is discussed in more detail in the following section and might involve providing visual aids, breaking tasks down into smaller steps, providing additional practice activities, or providing one-on-one support from a teacher, instructional aide, or peer mentor. By providing targeted support and remediation, you can ensure that all students have the necessary skills and knowledge to achieve the learning target.

You will also need to monitor your students' progress toward mastering prerequisite skills. This might involve using formative assessments to track student understanding of key concepts, providing frequent feedback on student work, or monitoring progress toward specific goals. You should be prepared to differentiate instruction and provide individualized support to meet the needs of each student.

You should provide opportunities for students to practice and reinforce prerequisite skills as they work towards achieving the larger goal. Similar to the setting and monitoring of a learning target, you can do this by following the steps below:

1. **Identify the prerequisite skills:** Begin by identifying the prerequisite skills that are necessary for students to understand the lesson material. This may include knowledge or skills that were previously taught or general cognitive abilities that are necessary for comprehension.

2. **Plan for review:** Once you have identified the prerequisite skills, plan to review them before introducing the new material. This

can be done through a variety of methods, such as a quick review activity or a pre-assessment to gauge student understanding.

3. **Provide scaffolding:** If some students are struggling with the prerequisite skills, provide additional support or scaffolding to help them succeed. This may include providing additional resources, such as videos or visual aids, or breaking down the material into smaller, more manageable steps.

4. **Differentiate instruction:** Consider how you can differentiate instruction to meet the needs of all learners, including those who may be struggling with prerequisite skills. This may include providing additional support—such as one-on-one instruction or modified assignments—or challenging activities for high-achieving students.

5. **Regularly assess student understanding:** Throughout the lesson, monitor and assess student understanding of the prerequisite skills. Use formative assessments—such as quick checks or exit tickets—to gather data on student understanding and adjust instruction as needed.

Add Scaffolding

Scaffolding refers to the supports and strategies that you use to help your students achieve the learning target. Scaffolding is important for students who may struggle with the material, as it can help to break down complex concepts into smaller, more manageable pieces. By identifying the scaffolding needed for each student, you can ensure that all students are able to make progress toward their goals.

There are many different types of scaffolding that you can use to support your students. Some examples include

- **Visual aids:** Visual aids can be helpful for students who struggle with reading or have difficulty processing information through text. This might include charts, graphs, diagrams, or pictures that help to illustrate key concepts.

- **Graphic organizers:** Graphic organizers are tools that help students to organize their thoughts and ideas. They can be particularly helpful for students who struggle with writing or have difficulty with executive functioning. Examples of graphic organizers include concept maps, Venn diagrams, and *know, wonder, learned* (KWL) charts.

- **Hands-on activities:** Hands-on activities can be an effective way to help students engage with the material and make connections between concepts. This might include experiments, simulations, or other activities that allow students to explore the material in a more concrete way.

- **Verbal prompts:** Verbal prompts can be helpful for students who struggle with attention or memory. This might include cues or reminders to help students stay on task or remember key information.

- **Modeling:** Modeling is a powerful way to demonstrate key concepts and skills. You can model the desired behavior or performance, allowing students to observe and practice the skill themselves.

When adding scaffolding to a lesson plan, consider the individual needs of each student. Some students may require more support than others, and you should be prepared to differentiate instruction to meet the needs of each student. This might involve providing additional visual aids, simplifying language, or breaking tasks down into smaller steps.

Scaffolding should be introduced gradually as students become more comfortable with the material. You should also be prepared to remove scaffolding as students become more independent and confident in their abilities.

Scaffolding should support, rather than detract from, the overall goals of the lesson. You should be intentional in your use of scaffolding and able to explain how each strategy is supporting student learning.

You can also incorporate scaffolding into your overall classroom routines and procedures. For example, you can provide visual schedules or checklists to help students stay on task and organize their work. You can also incorporate routines and procedures that promote independence, such as encouraging students to ask for help when they need it or providing opportunities for self-reflection and self-assessment. Here are some detailed instructions on how to add scaffolding to your lesson plan:

1. **Identify the complex tasks**: Begin by identifying the complex tasks in your lesson plan. These are the tasks that may be difficult for some students to complete independently.

2. **Break down the tasks:** Once you have identified the complex tasks, break them down into smaller, more manageable steps. This can be done by breaking the task into smaller sub-tasks or by providing students with a checklist of steps to follow.

3. **Provide guidance and support:** Provide guidance and support to students as they work through the scaffolded tasks. This may include additional resources or materials—such as graphic organizers or prompts—or one-on-one support.

4. **Gradually release responsibility:** As students become more comfortable with the scaffolded tasks, gradually release responsibility by providing less guidance and support. This will help students to develop their independent learning skills.

5. **Regularly assess student understanding:** Throughout the lesson, monitor and assess student understanding of the scaffolded tasks. Use formative assessments to gather data on student understanding and adjust instruction as needed.

Allow for Flexibility

Allowing for flexibility and accommodations is necessary for you to create effective lesson plans for your students. Accommodations refer to changes in the way that instruction is delivered, assignments are completed, or assessments are administered in order to make the curriculum accessible to students with diverse learning needs. By identifying and planning for these accommodations, you can help ensure that all students have the support they need to succeed.

Accommodations may take many forms, depending on the individual needs of each student. Some common accommodations include

- **Extended time:** Many special education students require additional time to complete assignments or assessments. Providing extended time allows students to work at their own pace and reduces the pressure and anxiety associated with time-limited tasks.

- **Modified assignments:** Some students may require modifications to assignments in order to make them more accessible. This might include simplifying instructions, reducing the amount of work required, or providing alternative formats— such as allowing a verbal presentation assignment to be submitted in portfolio format for an autistic student who struggles with public speaking.

- **Assistive technology:** Assistive technology can be a powerful tool for special education students. This might include text-to-speech software, screen readers, speech recognition software, or other tools that help students access and engage with the curriculum.

- **Alternative assessment formats:** Some students may require alternative assessment formats in order to demonstrate their understanding of the material. This might include oral exams, performance assessments, or alternative written assignments. Additional assessment accommodations might include, for instance, the provision of a reader or a scribe for students with dyslexia, vision impairment, or dexterity issues. However, options that provide for student independence and autonomy are always preferred.

- **Flexible seating and classroom arrangements:** For students with sensory processing issues or physical disabilities, flexible seating and classroom arrangements can be critical. This might include using stability balls instead of chairs, providing fidget toys, or rearranging the classroom to create more space.

Accommodations should be tailored to the specific needs of the student, rather than being applied uniformly across the class. You should work closely with parents and other members of each student's IEP team to identify appropriate accommodations and ensure that they are being implemented effectively. Accommodations should not fundamentally change the curriculum or make it easier for students to achieve the learning target. Instead, they should provide the necessary support for students to access the material and demonstrate their understanding while respecting the challenges or limitations that the student is facing.

Special education students often require different teaching strategies in order to access the curriculum. This might include using multisensory approaches, providing frequent breaks, or incorporating movement and physical activity into lessons. You should be willing to adapt your teaching to meet the individual needs of your students, rather than expecting all students to learn in the same way.

Remember that it's important to be flexible in terms of pacing and scheduling. Special education students may require more time to complete assignments or may need to take breaks more often than regular students. You should be willing to adjust their schedules to accommodate these needs, rather than expecting students to conform to a rigid schedule. Adding elastic content to lesson plans—which supports but is not essential to the learning target—can be a great way to ensure that you can vary the pacing of your classes to suit the day-to-day variations in student needs. For instance, a student with ASD might generally be a peer leader in your classroom, excelling through tasks and needing additional work to stay engaged. However, if they have had a sensorily difficult day, this pattern may reverse. In a case such as this, it is far more supportive of your student's needs to be able to remove some elastic content from your lesson plan and allow your classroom to be a safe space in which to self-regulate—for example, through stimming, using fidget toys, or wearing ear defenders—than to try to force that student to meet their usual achievement levels. By remaining flexible in this manner, you not only ensure student safety but also build trust and rapport with your students which will help them to engage and feel safe making mistakes around you in the future.

Every student is unique and may require different types of support in order to succeed. You should be willing to experiment with different accommodations and instructional approaches in order to find what works best for each student.

Include Modifications and Differentiation

Modifications refer to changes in the curriculum or instruction that are made in order to make the material more accessible to students with diverse learning needs. Differentiation refers to varying the content, process, or product of instruction to meet the individual needs of students.

There are many different ways to incorporate modifications and differentiation into lesson plans. Some strategies include

- **Modifying the curriculum:** In some cases, it may be necessary to modify the curriculum to make it more accessible to special education students. This might involve simplifying the content, breaking it down into smaller chunks, or providing additional examples or explanations.

- **Adjusting the pacing:** As discussed in the previous section, special education students may require more time to process information or complete assignments. Adjusting the pacing of the lesson to allow for more time or providing opportunities for review and reinforcement can help ensure that students are able to keep up with the material.

- **Providing alternative assignments:** For some students, the standard assignments may be too challenging or not well-suited to their learning needs. Providing alternative assignments that allow students to demonstrate their understanding of the material in a different way can help ensure that all students are able to succeed.

- **Varying instructional strategies:** Different students may respond to different types of instruction. Varying instructional strategies—such as using multisensory approaches or incorporating hands-on activities—can help engage students and ensure that they are able to access the material.

- **Providing additional support:** Some students may require additional support in order to succeed. This might include providing one-on-one support from a teacher or using peer tutoring or collaborative learning strategies.

Modifications and differentiation should be tailored to the specific needs of each student, rather than being applied uniformly across the class.

You should work closely with parents and other members of the student's IEP team to identify appropriate modifications and ensure that they are being implemented effectively.

You will also need to create a supportive and inclusive classroom environment. This might involve using positive reinforcement, providing opportunities for peer collaboration, or incorporating culturally responsive teaching practices. You should be willing to experiment with different approaches and strategies in order to find what works best for each student.

Incorporate Technology

As we discussed earlier, with the rise of educational technology, you now have access to a wide range of tools and resources that can be used to enhance instruction and support student success. Some ways that you can include technology in your lesson plans can be through the use of

- **Educational apps:** There are a wide range of educational apps available for tablets and smartphones that can help reinforce key concepts and skills. These apps can be particularly useful for students who struggle with traditional teaching methods or who benefit from interactive, hands-on learning experiences. See Chapter 4 for a list of suggested apps to use in your classroom.

- **Interactive whiteboards:** Interactive whiteboards allow you to create dynamic, engaging lesson plans that can be tailored to the individual needs of each student. You can use interactive whiteboards to display multimedia content, create virtual simulations, and facilitate collaborative learning activities.

- **Assistive technology:** Assistive technology refers to tools and devices that are designed to help students with disabilities access the curriculum and participate in classroom activities. Examples of assistive technology include screen readers, text-to-speech software, and speech recognition software.

- **Online learning platforms:** Online learning platforms such as *Google Classroom* or *Moodle* can be used to provide students with access to digital resources, facilitate online discussions, and provide feedback on assignments. These platforms can be particularly useful for students who struggle with traditional classroom environments or who require more flexible learning opportunities.

Not all students will benefit from the same types of technology, and some may require additional support in order to use technology effectively. Technology should be used as a tool to support student learning, rather than as a substitute for traditional teaching methods. Strike a balance between traditional teaching methods and technology-based instruction to ensure that all students are able to access the material and achieve their full potential.

Remember to provide students with opportunities to develop their digital literacy skills. This might involve providing instruction on how to use different types of software or devices or creating opportunities for students to collaborate and problem-solve using technology.

Decide How the Material Will Be Presented

There are many different ways to present each topic, and the best method for your class will depend on the individual needs and learning styles of each student.

Some strategies for presenting material that you should consider include

- **Visual aids:** Visual aids can be an effective way to communicate information to students who are visual learners. Examples of visual aids might include charts, graphs, diagrams, or photographs. Visual aids can help students understand complex concepts and remember key information.

- **Auditory prompts:** For students who are auditory learners, auditory prompts can be an effective way to communicate information. Examples of auditory prompts might include verbal instructions, recorded lectures, songs, or podcasts. Auditory prompts can help students focus on key information and remember important concepts.

- **Hands-on activities:** Hands-on activities can be an effective way to engage students who are experiential learners and help them understand abstract concepts. Examples of hands-on activities might include experiments, poster-creation projects, or role-playing activities. Hands-on activities can help students connect the material to real-life situations and develop their problem-solving and critical-thinking skills.

- **Technology-based instruction:** As discussed in the previous section, technology can be an effective way to present material to students, particularly those who struggle with traditional teaching methods. Examples of technology-based instruction might include educational apps, interactive whiteboards, or online learning platforms.

Some students may benefit from a combination of visual aids and auditory prompts, while others may require more hands-on activities to stay engaged. Breaking up material into smaller, manageable chunks can help students stay focused and retain information. Providing opportunities for review and repetition can also be helpful for students who struggle with memory and recall.

Think about how the material fits into the larger curriculum. You should work to connect the material to real-life situations and provide opportunities for students to apply what they've learned in different contexts.

Create a Guide to Complete Each Activity

You should develop a step-by-step guide for each activity. This guide should provide clear instructions, a list of materials needed, and any necessary supports or accommodations to help students successfully complete the activity.

Here is how you can go about creating a guide for each activity in your lesson plan:

1. **Identify the learning objectives:** Before creating the guide, it's important to identify the specific learning objectives for each activity. What skills or knowledge do you want your students to gain from this activity?

2. **Break down the steps:** Once you've identified the learning objectives, break down the activity into smaller, manageable steps. This will help students stay focused and understand what is expected of them.

3. **Provide clear instructions:** Write clear, concise instructions for each step of the activity. Use simple language, and avoid complex sentences or technical jargon. Consider using visual aids or diagrams to help illustrate the steps.

4. **Identify necessary materials:** Identify all of the materials and supplies needed for the activity. This may include books, worksheets, art supplies, or technology tools. Be sure to include any special accommodations or modifications that may be needed, such as larger print materials or adaptive technology.

5. **Plan for supports and accommodations:** Consider the individual needs of each student and plan for any necessary supports or accommodations. This may include extra time, simplified instructions, or additional support from a teacher or aide.

6. **Incorporate assessment and feedback:** Plan for how you will assess students' understanding of the activity and provide feedback. This may include informal observation, formative assessments, or summative assessments.

7. **Review and revise:** After creating the guide, review it carefully to ensure that it is clear and easy to follow. Revise as necessary to make sure that it meets the needs of all students.

Decide How Progress Will Be Monitored and Assessed

When planning a lesson, you'll need to think about how you will monitor and assess student progress throughout the lesson and beyond. Here are some things to keep in mind:

1. **Determine what data to collect:** Before the lesson begins, identify what data you will collect to measure student progress toward the learning objective. This could include informal observations, work samples, quizzes, or other assessments. Be sure to consider what types of data are appropriate for each student's individualized needs and goals.

2. **Decide how often to collect data:** Depending on the length and complexity of the lesson, you may need to collect data at different intervals. Consider collecting data at the beginning and end of the lesson, as well as at various checkpoints throughout. This will give you a more comprehensive picture of student progress and help you make informed instructional decisions.

3. **Use data to make instructional decisions:** Once you've collected the data, use it to make informed instructional decisions. Analyze the data to identify areas of strength and areas for improvement, and adjust your instruction accordingly. This might mean providing additional supports or accommodations, modifying the curriculum, or adjusting the pacing of the lesson.

4. **Involve students in the assessment process:** Involving students in the assessment process can help them take ownership of their learning and increase their motivation. Consider using self-assessment tools, peer evaluations, or other student-centered assessment strategies. Be sure to provide clear guidance and support to ensure that students are able to effectively evaluate their own progress.

5. **Communicate progress to families and other stakeholders:** Regular communication is essential to ensuring that everyone is aware of student progress and can work collaboratively to support student learning. Be sure to provide regular updates on student progress, communicate any changes to the lesson or supports being provided, and solicit feedback and input from families and other stakeholders. This will help ensure that everyone is on the same page and working towards the same goals.

There are a couple of ways that you can monitor and assess student progress in the classroom. Here are some examples:

- **Observations:** Observations are a great way to assess student progress and gather information about how they are performing. You can observe students as they complete tasks or engage in activities and take note of their behaviors, interactions, and understanding of the material.

- **Work samples:** Work samples can provide a tangible record of student progress and demonstrate how they are applying what they have learned. You can collect and analyze work samples to identify areas of strength and weakness and adjust instruction accordingly.

- **Quizzes and tests:** Quizzes and tests can be used to assess student understanding of key concepts and skills. You can use these assessments to identify areas where students need additional support.

- **Self-assessment tools:** Self-assessment tools, such as checklists or rubrics, can help students evaluate their own progress and take ownership of their learning. You can provide guidance and support to help students effectively use these tools.

- **Peer evaluations:** Peer evaluations can provide valuable feedback to students and help them identify areas where they need additional support. You can provide clear guidance and support to ensure that students are able to effectively evaluate their peers.

- **Progress monitoring tools:** Progress monitoring tools can provide a structured way to collect data and track student progress over time. You can use these tools to identify trends and patterns in student performance and adjust instruction if needed.

- **Parent-teacher conferences:** Parent-teacher conferences can provide an opportunity for you to share information about student progress and solicit feedback and input from parents. These conferences can be used to set goals, identify areas of concern, and develop plans for ongoing support.

Include Necessary Materials

This includes not only traditional classroom supplies, such as pencils and paper, but also any special materials or accommodations that may be required to support individual students' needs. To ensure that you have all the necessary materials for your lesson plans you should

- **Review your lesson plan:** Before you begin gathering materials, review your lesson plan carefully to ensure that you have a clear understanding of what will be required for each activity. This will help you identify any unique materials or accommodations that may be needed.

- **Consider individual student needs:** Each student has unique needs and preferences, so it is important to consider these when gathering materials. For example, students with visual impairments may require braille materials, while those with mobility impairments may need specialized seating or equipment.

- **Make a list:** Once you have a clear understanding of what will be required, make a list of all the materials you will need for each activity. This will help you stay organized and ensure that you do not overlook any important items.

- **Gather materials in advance:** It is important to gather all necessary materials in advance of each lesson. This will help you avoid disruptions or delays during class and ensure that you have time to make any necessary accommodations or modifications.

- **Store materials in an organized manner:** Once you have gathered all necessary materials, it is important to store them in a way that is organized and easily accessible. This will help you locate materials quickly during class and ensure that they are not lost or damaged.

- **Consider sensory preferences:** Many special education students have unique sensory preferences, such as a preference for tactile or visual materials. When gathering materials, consider these preferences and try to incorporate materials that will engage students in a meaningful way.

- **Make accommodations as needed:** Special education students often require accommodations to access the curriculum. When gathering materials, be sure to consider any accommodations that may be needed, such as larger print materials or colored filter sheets.

Make Sure the Plan Is Aligned With Set Standards

When creating a lesson plan for special education students, make sure that the plan aligns with the set academic standards for the grade level and subject area. This helps ensure that students are receiving instruction that is on par with their peers and that they are making progress toward their academic goals.

You can do this by

1. **Determining the relevant academic standards:** Each state has its own set of academic standards for each grade level and subject area. These standards outline the knowledge and skills that students should master by the end of the school year. You can access the state standards online or through their district's curriculum department.

2. **Reviewing the standards carefully:** Once you have identified the relevant standards, review them carefully to determine what knowledge and skills the students should have by the end of the lesson or unit. This will help you determine what content to include in your lesson and what activities and assessments to use to measure student learning.

3. **Aligning lesson objectives with the standards:** The objectives of the lesson should align with the standards. This means that the objectives should be specific, measurable, and attainable, and they should address the knowledge and skills outlined in the standards. For example, if the standard is for students to be able to identify the main objective of any age-appropriate word problem, the objective of the lesson might be for students to learn and use operations vocabulary.

4. **Using the standards to guide instructional decisions:** The standards can guide decisions such as what materials to use and what instructional strategies to employ. For example, if the standard requires students to be able to write an explanation of a mathematical proof, the teacher might select a text to use as a model, provide graphic organizers to help students plan their essays, and provide opportunities for students to practice writing their essays before completing a final draft.

5. **Aligning assessments with the standards:** Design assessments that measure the knowledge and skills outlined in the standards. This might involve using multiple types of assessments, such as quizzes, tests, and projects. The assessments should be designed to measure student learning and help you identify areas where students need additional support.

6. **Making adjustments as necessary:** As the lesson progresses, monitor student progress and adjust the plan as necessary to ensure that students are making progress towards the standards. This might involve re-teaching certain concepts, providing additional supports or accommodations, or adjusting the pacing of the lesson.

Blank Lesson Plan

Here is a blank lesson plan that you can use or modify.

Lesson Title:

Grade Level:

Subject Area:

Lesson Duration:

Learning Target(s):

Prerequisite Skills:

Materials Needed:

Accommodations Needed:

Modifications/Differentiation:

Technology Integration:

Presentation of Material:

Scaffolding:

Step-by-Step Guide for Each Activity:

Assessment Plan:

Alignment with Academic Standards:

Reflection/Evaluation:

Note: As the lesson progresses, the teacher should document student progress and adjust the lesson plan as needed to ensure that all students are achieving the learning target(s).

Tying It All Together

Teaching can be overwhelming, especially when it comes to planning lessons. As a teacher, it is easy to get lost in the details and lose sight of the big picture. Creating a visual plan is a powerful tool that can help you to tie everything together and maintain a sense of focus and direction.

Try using large pieces of paper and different colored sticky notes. This technique allows you to organize your lesson plans in a way that is visually appealing and easy to understand.

Choosing a color theme is the first step in creating a visual plan. Different colors can be used for different subjects or concepts. For example, multiplication concepts can be assigned one color, while fraction concepts can be assigned another. This makes it easy to quickly identify the main ideas and concepts of each lesson.

Next, you can begin to use sticky notes to plan out your lessons. Each sticky note can represent one lesson, and they can be organized on a piece of paper to create a long-term plan. This plan can be organized by week, month, or term, depending on your preference.

Drawing a calendar on each piece of paper is a helpful way to structure the lesson plans. You can use the calendar to identify important dates, such as holidays or exams, and plan lessons accordingly. The sticky notes can then be placed in the lesson slots to create an organized and easy-to-follow plan.

If you realize that a particular lesson needs to be rearranged, you can simply move the sticky note to a different slot on the calendar. This allows for a level of flexibility and adaptability that is essential in the world of education.

A visual plan can help you identify gaps or overlaps in your lesson plans. By looking at the plan as a whole, you can ensure that you are covering all of the necessary topics and concepts. You can also identify times when you may be spending too much time on one topic and not enough on another. By sharing the plan with students, parents, or colleagues, you can provide a clear and concise overview of what will be covered in class. This can help students stay organized and focused and can also help parents stay informed about what their child is learning.

Creating a visual plan does not have to be a time-consuming or complicated process. With just a few pieces of paper and some sticky notes, you can create a plan that is easy to understand and flexible enough to adapt to changing circumstances. By taking the time to create a visual plan, you can save time and stress in the long run.

Where to Find Fun, Engaging Resources

When it comes to finding resources for students with disabilities, there are many websites and online platforms that offer rich materials, manipulatives, worksheets, and activities. Here are some of the best resources available:

- **Against All Odds—Inside Statistics:** This website offers a comprehensive series of videos and activities designed to teach statistics to high school students. The site includes resources for students with disabilities, including closed captioning, audio descriptions, and downloadable transcripts.
- **Eedi:** This site offers free resources and support for math students of all ages, including those with disabilities. It includes

a range of interactive activities, practice exercises, and downloadable worksheets.

- **GeoGebra:** This site offers a range of interactive tools and activities designed to teach math concepts to students of all ages. It includes resources for students with disabilities, including accessibility features such as text-to-speech and screen reader support.

- **SolveMe Puzzles:** This site offers a collection of interactive puzzles and challenges designed to teach problem-solving and critical-thinking skills. It includes resources for students with disabilities, including support for screen readers and alternative input methods.

- **PhET Simulations:** This site offers a collection of interactive simulations designed to teach physics, chemistry, math, and other STEM subjects. It includes resources for students with disabilities, including accessibility features such as closed captioning and adjustable font sizes.

- **Yummy Math:** This site offers a range of math activities and lessons that are designed to be engaging and relevant to students' lives. It includes resources for students with disabilities, including materials that are available in different formats and levels of difficulty.

- **Brilliant:** This site offers a collection of interactive courses designed to teach math concepts to students of all ages. It includes resources for students with disabilities, including accessibility features such as audio descriptions and alternative input methods.

- **eMathInstruction:** This site offers a range of instructional materials and resources designed to teach math concepts to students of all ages. It includes resources for students with disabilities, including materials that are available in different formats and levels of difficulty.

- **WeAreYou:** This site offers a collection of the best math websites and resources available for students of all ages. It includes resources for students with disabilities, including materials that are accessible and available in different formats.

- **IXL Learning:** This site offers a collection of interactive practice exercises and activities designed to teach math and other subjects to students of all ages. It includes resources for students with disabilities, including materials that are available in different formats and levels of difficulty.

Choosing to be a special needs educator requires an extra large heart full of compassion and empathy. To be an effective and confident special needs teacher you need a plan! Taking time for detailed planning will free up time later on, time that will be better spent being present with students!

Conclusion

Math really has gained a bad reputation, so it's no wonder that special needs students—whether with physical disabilities, neurodivergence, learning delays, or emotional disorders—are going to have a much harder time with the subject. Once a student starts to lose confidence and motivation, it's a slippery downhill slope. That is unless there is an awesome teacher who is ready to improve their teaching strategies.

Teaching math to students who have given up on learning can be a challenging but rewarding experience. By understanding the unique needs and learning styles of these students, educators can develop effective strategies that foster engagement, build confidence, and ultimately lead to academic success.

Throughout this book, we have explored a range of evidence-based practices and techniques for teaching math to disengaged learners. We have discussed the importance of creating a safe and supportive learning environment, establishing clear goals and expectations, and providing students with personalized instruction and feedback. We have also highlighted the value of incorporating real-world applications and hands-on activities into math lessons, as well as the benefits of using technology and collaborative learning approaches. By adopting these strategies, you can make a meaningful difference in the lives of students who have given up on learning math. However, implementing all 11 strategies at once can be overwhelming, so it's best to start with one strategy at a time.

Ultimately, the key to success in teaching math to disengaged learners lies in the ability to connect with students on a personal level, inspire them to believe in their own abilities, and help them to see the relevance and importance of math in their daily lives. By adopting the strategies and techniques outlined in this book, you can make a meaningful difference in the lives of these students, setting them on a path to academic success and a brighter future.

There is a lot of support for teaching neurodiverse students. There are plenty of ideas for teaching high school students. There are endless resources for teaching math. Yet, there are very few resources that combine all three, which leaves you struggling to find effective strategies. So, if you've found this book to be a valuable resource, please consider leaving a review on Amazon. Your review can help others struggling to find effective strategies for teaching math to neurodivergent and disabled high school students. By sharing your opinions, you can help spread the word that there are resources available to support struggling students and inspire more teachers to adopt these effective strategies. Your review can make a real difference in changing math lessons for even more struggling students. Thank you for your support!

About the Author

Jordan B. Smith Jr. is an accomplished individual with a distinguished background in military leadership, education, and public service. Born and raised in St. Louis, Missouri, he attended Lexington Grade School before being selected to attend the prestigious Christian Brothers College Military Institute in Clayton, Missouri, where he achieved the Cadet Lieutenant Colonel rank, becoming the first African American to do so. Upon graduation, he received the Damian Saber Award for Military Leadership—another first for an African American—and was nominated to attend the United States Naval Academy, where he became the first African American selected as the 17th Color Company Commander in June of 1976.

After graduation from the Naval Academy, he served 20 years as a United States Marine, reaching the rank of Major as a Logistics Officer, and served in the Gulf War (1990–1991) with Marine Air Group 26 (MAG-26). He then transitioned to education and has been a public school mathematics teacher for over 19 years. Dr. Smith has earned multiple degrees, including a Doctorate in Educational Leadership with a specialization in curriculum and instruction, and he has been recognized as a CDE Model school field expert and a congressional advocate. He is also an active chairperson for Accrediting Commission for Schools Western Association of Schools and Colleges (ACS WASC), a school site WASC Coordinator, and an author, public speaker, conference presenter, and educational consultant.

Dr. Smith's unique talent for finding multiple ways to motivate at-risk students and build their confidence in learning math concepts has contributed significantly to increased graduation rates on the campus where he has worked for the past 10 years. His dedication to education and public service has also led to numerous achievements, including the recruitment of graduating students to serve our country in the Armed

Forces of the United States. In his free time, he enjoys singing and music production.

Glossary

Arithmetic: The branch of mathematics that deals with basic operations such as addition, subtraction, multiplication, and division.

Algorithm: A step-by-step procedure for solving a mathematical problem.

Area: The measure of the surface enclosed by a two-dimensional shape.

Calculus: A branch of mathematics that deals with the study of rates of change and how things change over time.

Coordinate plane: A two-dimensional graph with an x-axis and a y-axis used to represent ordered pairs of numbers.

Decimal: A number expressed in the base-10 system, using a decimal point to indicate the place value of each digit.

Equation: A mathematical statement that uses symbols, numbers, and operations to express a relationship between two or more quantities.

Fraction: A number that represents a part of a whole, expressed as one number divided by another.

Geometry: The branch of mathematics that deals with the study of shapes, sizes, and spatial relationships.

Graph: A visual representation of data or a mathematical function.

Integer: A whole number, either positive, negative, or zero, without a fractional or decimal component.

Measurement: The process of determining the size, length, or amount of something.

Mean: The average value of a set of numbers, obtained by adding them together and dividing by the number of values.

Median: The middle value of a set of numbers, obtained by arranging them in numerical order and finding the value that divides the set in half.

Mode: The value that occurs most frequently in a set of numbers.

Number line: A line used to represent the set of real numbers, with negative numbers to the left of zero and positive numbers to the right.

Operations: Mathematical processes such as addition, subtraction, multiplication, and division.

Perimeter: The distance around the edge of a two-dimensional shape.

Probability: The likelihood or chance of a particular event occurring.

Ratio: A comparison of two quantities, expressed as a fraction or using a colon.

Statistics: The collection, analysis, interpretation, presentation, and organization of data.

Variable: A symbol or letter used to represent a quantity or value that can change in a mathematical expression or equation.

Volume: The amount of space enclosed by a three-dimensional object.

X-axis: The horizontal axis on a coordinate plane.

Y-axis: The vertical axis on a coordinate plane.

References

Against all odds: Inside statistics. (2020, March 19). Annenberg Learner. https://www.learner.org/series/against-all-odds-inside-statistics/

Alegre-Ansuategui, F,J. Moliner, L. Lorenzo, G. Maroto, A.(2017, August 30). *Peer Tutoring and Academic Achievement in Mathematics: A Meta- Analysis.*Eurasia Journal of Mathematics, Science and Technology Education. Retrieved March 22, 2023, from https://www.ejmste.com/download/peer-tutoring-and-academic-achievement-in-mathematics-a-meta-analysis-5265.pdf

Armstrong, T. (2013, April 9). *7 Ways to Bring Out the Best in Special-Needs Students (Opinion).* Education Week. Retrieved March 8, 2023, from https://www.edweek.org/teaching-learning/opinion-7-ways-to-bring-out-the-best-in-special-needs-students/2013/04

Bhalla, M. (2020, July 4). *Special Education v/s Inclusive Education: How are they similar or different?* Mahima Bhalla. Retrieved March 6, 2023, from https://mahimabhalla.medium.com/special-education-v-s-inclusive-education-how-are-they-similar-or-different-3f1127a013a8

Covid-19: School closures - free maths support, resources and ideas. (n.d.). Eedi. https://eedi.com/blog/covid-19-school-closures-free-maths-support-resources-and-ideas

Dubin, A. (2022, March 10). *101 Best Math Jokes for Kids - Funny Math Puns for Kids and Students.* Woman's Day. Retrieved March 8, 2023, from https://www.womansday.com/life/entertainment/a39124261/math-jokes/

eMath Homepage. (2022, July 14). eMATHinstruction. https://www.emathinstruction.com/

Feynman, R. (2022, October 2). *Top 100 Brilliant Math Quotes To Inspire Students and Teachers.* Quote.cc. Retrieved March 6, 2023, from https://www.quote.cc/math-quotes/

Fink, S. (2020, February 11). *Benefits of Peer Teaching · Coditum.* SummerTech. Retrieved March 22, 2023, from https://www.summertech.net/benefits-of-peer-teaching/

Fisher, D., & Frey, N. (2013). The CCSS and mathematics vocabulary: Frayer model revisited. Mathematics Teaching in the Middle School, 19(1), 6-11.

Gallo, M. A., & McDougall, D. (2013). Using the Frayer model to promote conceptual understanding in mathematics. Mathematics Teaching in the Middle School, 19(4), 224-230.Mooney, E. S., & Maheady, L. (2014). Enhancing mathematical problem solving of fourth-grade students with learning disabilities: The Frayer model. Learning Disabilities: A Contemporary Journal, 12(1), 19-29.

Griffith, R. (n.d.). Funny Joke to Tell. Retrieved March 23, 2023, from https://www.pinterest.es/pin/19703317104643115/

Ismail, M. (2020, October 22). *Strategies for Motivating Students: Start with Intrinsic Motivation.* Waterford.org. Retrieved March 8, 2023, from https://www.waterford.org/education/how-to-motivate-students/

IXL: Math, Language Arts, Science, Social Studies, and Spanish. (n.d.). IXL Learning. https://www.ixl.com/

Jenkins, C., & Bofah, Y. (2022, August 19). *69 Best Friendship Quotes - Meaningful Sayings About True Friends.* Good Housekeeping.

Retrieved March 22, 2023, from https://www.goodhousekeeping.com/life/relationships/g5055/friendship-quotes/

Joke of the week. (2021, May 3). Daventry Library. *Facebook.* Retrieved March 23, 2023, from https://www.facebook.com/daventrylibrary/photos/a.10151034942696168/10159473339171168/?type=3&paipv=0&eav=AfYkfnGVRrYTiXoy3rCE5TMT5DLltFpGdQ5FEoMxx5575uSkZGjcftkYno7RDL9JpYA&_rdr

Learn interactively - Math. (n.d.). Brilliant. https://brilliant.org/landing/interactive-courses-learn-math/

Markham, D. B. (2021, June 22). *Math Dad Jokes.* Nerd Ducks. Retrieved March 22, 2023, from https://danielbmarkham.com/math-dad-jokes/

Math & Movement. (2015, June 6). Math & Movement: Movement-Based Learning for Your School. https://mathandmovement.com/

Math Jokes. (2016, May 2). dpsmaths. Retrieved March 23, 2023, from http://dpsmaths.weebly.com/Charles, R., & Lester, F. K. (1982). An instructional model for using learning theory in mathematics education. Educational Studies in Mathematics, 13(2), 171 185.

McKinney, A. (2020, May 3). *The Importance of Building Community in the Classroom.* TeacherVision. Retrieved March 6, 2023, from https://www.teachervision.com/blog/morning-announcements/importance-building-community-classroom

Middle & High S. Math: 2200+ resources. (2020, April 14). GeoGebra. from https://www.geogebra.org/m/kewpjrue

Motivating Special Needs Children. (2019, May 7). Special Learning. Retrieved March 8, 2023, from https://special-learning.com/article/motivating-special-needs-children/

PhET Interactive Simulations. (n.d.). PhET. https://phet.colorado.edu/

Posamentier, A. (2020, May 6). *9 Strategies for Motivating Students in Mathematics.* Edutopia. Retrieved March 8, 2023, from https://www.edutopia.org/blog/9-strategies-motivating-students-mathematics-alfred-posamentier

Singh, M. (2020, April 27). *Top 10 Math Manipulatives For High School Students.* Number Dyslexia. Retrieved March 6, 2023, from https://numberdyslexia.com/top-10-math-manipulatives-for-high-school-students/

SolveMe puzzles. (n.d.). SolveMe Puzzles. https://solveme.edc.org/

Special Education Classroom Setup: Designing a Structured Space for Special Needs. (2020, May 14). FishyRobb. Retrieved March 6, 2023, from https://www.fishyrobb.com/post/special-education-classroom-setup

25 Fun Math Activities for Middle & High School Students | Creative Club Exercises. (2021, February 19). iD Tech. Retrieved March 8, 2023, from https://www.idtech.com/blog/ways-to-make-math-more-fun-engaging

We are you - Best math websites. (n.d.). We are you. https://www.weareyou.com/best-math-websites/

Yummy Math. (n.d.). YummyMath. https://www.yummymath.com/

Image References

Allison, S. (2017, December 11). *Making Spaced Practice Count | Class Teaching*. Class Teaching. Retrieved March 23, 2023, from https://classteaching.wordpress.com/2017/12/11/making-spaced-practice-count/

Amanda. (2021, July 13). *The forgetting curve*. Organising Students. Retrieved March 23, 2023, from https://organisingstudents.com.au/2021/07/the-forgetting-curve/

All other images have been provided by the author.

www.ingramcontent.com/pod-product-compliance
Ingram Content Group UK Ltd.
Pitfield, Milton Keynes, MK11 3LW, UK
UKHW041428180426
11947UKWH00007B/339